Functional Data Structures in R

Advanced Statistical Programming in R

Thomas Mailund

Apress®

Functional Data Structures in R: Advanced Statistical Programming in R

Thomas Mailund
Aarhus N, Denmark

ISBN-13 (pbk): 978-1-4842-3143-2   ISBN-13 (electronic): 978-1-4842-3144-9
https://doi.org/10.1007/978-1-4842-3144-9

Library of Congress Control Number: 2017960831

Cover image by Freepik (www.freepik.com)

Managing Director: Welmoed Spahr
Editorial Director: Todd Green
Acquisitions Editor: Steve Anglin
Development Editor: Matthew Moodie
Technical Reviewer: Karthik Ramasubramanian
Coordinating Editor: Mark Powers
Copy Editor: Corbin P Collins

Distributed to the book trade worldwide by Springer Science+Business Media New York, 233 Spring Street, 6th Floor, New York, NY 10013. Phone 1-800-SPRINGER, fax (201) 348-4505, e-mail orders-ny@springer-sbm.com, or visit www.springeronline.com. Apress Media, LLC is a California LLC and the sole member (owner) is Springer Science + Business Media Finance Inc (SSBM Finance Inc). SSBM Finance Inc is a **Delaware** corporation.

For information on translations, please e-mail rights@apress.com, or visit www.apress.com/rights-permissions.

Apress titles may be purchased in bulk for academic, corporate, or promotional use. eBook versions and licenses are also available for most titles. For more information, reference our Print and eBook Bulk Sales web page at www.apress.com/bulk-sales.

Any source code or other supplementary material referenced by the author in this book is available to readers on GitHub via the book's product page, located at www.apress.com/9781484231432. For more detailed information, please visit www.apress.com/source-code.

Printed on acid-free paper

Table of Contents

About the Author

Thomas Mailund is an associate professor in bioinformatics at Aarhus University, Denmark. He has a background in math and computer science. For the last decade, his main focus has been on genetics and evolutionary studies, particularly comparative genomics, speciation, and gene flow between emerging species. He has published *Beginning Data Science in R*, *Functional Programming in R*, and *Metaprogramming in R* with Apress, as well as other books.

About the Technical Reviewer

Karthik Ramasubramanian works for one of the largest and fastest-growing technology unicorns in India, Hike Messenger, where he brings the best of business analytics and data science experience to his role. In his seven years of research and industry experience, he has worked on cross-industry data science problems in retail, e-commerce, and technology, developing and prototyping data-driven solutions. In his previous role at Snapdeal, one of the largest e-commerce retailers in India, he was leading core statistical modeling initiatives for customer growth and pricing analytics. Prior to Snapdeal, he was part of the central database team, managing the data warehouses for global business applications of Reckitt Benckiser (RB). He has vast experience working with scalable machine learning solutions for industry, including sophisticated graph network and self-learning neural networks. He has a master's degree in theoretical computer science from PSG College of Technology, Anna University, and is a certified big data professional. He is passionate about teaching and mentoring future data scientists through different online and public forums. He enjoys writing poems in his leisure time and is an avid traveler.

Introduction

This book gives an introduction to functional data structures. Many traditional data structures rely on the structures being mutable. We can update search trees, change links in linked lists, and rearrange values in a vector. In functional languages, and as a general rule in the R programming language, data is not mutable. You cannot alter existing data. The techniques used to modify data structures to give us efficient building blocks for algorithmic programming cannot be used.

There are workarounds for this. R is not a pure functional language, and we can change variable-value bindings by modifying environments. We can exploit this to emulate pointers and implement traditional data structures this way; or we can abandon pure R programming and implement data structures in C/C++ with some wrapper code so we can use them in our R programs. Both solutions allow us to use traditional data structures, but the former gives us very untraditional R code, and the latter has no use for those not familiar with other languages than R.

The good news, though, is that we don't have to reject R when implementing data structures if we are willing to abandon the traditional data structures instead. There are data structures that we can manipulate by building new versions of them rather than modifying them. These data structures, so-called *functional data structures*, are different from the traditional data structures you might know, but they are worth knowing if you plan to do serious algorithmic programming in a functional language such as R.

There are not necessarily drop-in replacements for all the data structures you are used to, at least not with the same runtime performance for their operations, but there are likely to be implementations for most abstract data structures you regularly use. In cases where you might have to lose a bit of efficiency by using a functional data structures instead of a traditional one, however, you have to consider whether the extra speed is worth the extra time you have to spend implementing a data structure in exotic R or in an entirely different language.

There is always a trade-off when it comes to speed. How much programming time is a speed-up worth? If you are programming in R, chances are you value programmer-time over computer-time. R is a high-level language and relatively slow compared to most other languages. There is a price to providing higher levels of expressiveness. You accept this when you choose to work with R. You might have to make the same choice when it comes to selecting a functional data structure over a traditional one, or you might conclude that you really do need the extra speed and choose to spend more time programming to save time when doing an analysis. Only you can make the right choice based on your situation. You need to find out the available choices to enable you to work data structures when you cannot modify them.

CHAPTER 1

Introduction

This book gives an introduction to functional data structures. Many traditional data structures rely on the structures being mutable. We can update search trees, change links in linked lists, and rearrange values in a vector. In functional languages, and as a general rule in the R programming language, data is *not* mutable. You cannot alter existing data. The techniques used to modify data structures to give us efficient building blocks for algorithmic programming cannot be used.

There are workarounds for this. R is not a *pure* functional language, and we can change variable-value bindings by modifying environments. We can exploit this to emulate pointers and implement traditional data structures this way; or we can abandon pure R programming and implement data structures in C/C++ with some wrapper code so we can use them in our R programs. Both solutions allow us to use traditional data structures, but the former gives us very untraditional R code, and the latter has no use for those not familiar with other languages than R.

The good news, however, is that we don't have to reject R when implementing data structures if we are willing to abandon the traditional data structures instead. There are data structures we can manipulate by building new versions of them rather than modifying them. These data structures, so-called *functional data structures*, are different from the traditional data structures you might know, but they are worth knowing if you plan to do serious algorithmic programming in a functional language such as R.

T. Mailund, *Functional Data Structures in R*, https://doi.org/10.1007/978-1-4842-3144-9_1

There are not necessarily drop-in replacements for all the data structures you are used to, at least not with the same runtime performance for their operations—but there are likely to be implementations for most abstract data structures you regularly use. In cases where you might have to lose a bit of efficiency by using a functional data structure instead of a traditional one, you have to consider whether the extra speed is worth the extra time you have to spend implementing a data structure in exotic R or in an entirely different language.

There is always a trade-off when it comes to speed. How much programming time is a speed-up worth? If you are programming in R, the chances are that you value programmer time over computer time. R is a high-level language that is relatively slow compared to most other languages. There is a price to providing higher levels of expressiveness. You accept this when you choose to work with R. You might have to make the same choice when it comes to selecting a functional data structure over a traditional one, or you might conclude that you really *do* need the extra speed and choose to spend more time programming to save time when doing an analysis. Only you can make the right choice based on your situation. You need to find out the available choices to enable you to work data structures when you cannot modify them.

CHAPTER 2

Abstract Data Structures

Before we get started with the actual data structures, we need to get some terminologies and notations in place. We need to agree on what an *abstract* data structure is—in contrast to a *concrete* one—and we need to agree on how to reason with runtime complexity in an abstract way.

If you are at all familiar with algorithms and data structures, you can skim quickly through this chapter. There won't be any theory you are not already familiar with. Do at least skim through it, though, just to make sure we agree on the notation I will use in the remainder of the book.

If you are not familiar with the material in this chapter, I urge you to find a text book on algorithms and read it. The material I cover in this chapter should suffice for the theory we will need in this book, but there is a lot more to data structures and complexity than I can possibly cover in a single chapter. Most good textbooks on algorithms will teach you a lot more, so if this book is of interest, you should not find any difficulties in continuing your studies.

© Thomas Mailund 2017
T. Mailund, *Functional Data Structures in R*, https://doi.org/10.1007/978-1-4842-3144-9_2

Structure on Data

As the name implies, data structures have something to do with structured data. By *data*, we can just think of elements from some arbitrary set. There might be some more structure to the data than the individual data points, and when there is we keep that in mind and will probably want to exploit that somehow. However, in the most general terms, we just have some large set of data points.

So, a simple example of working with data would be imagining we have this set of possible values—say, all possible names of students at a university—and I am interested in a subset—for example, the students that are taking one of my classes. A *class* would be a subset of students, and I could represent it as the subset of student names. When I get an email from a student, I might be interested in figuring out if it is from one of *my* students, and in that case, in which class. So, already we have some structure on the data. Different classes are different subsets of student names. We also have an operation we would like to be able to perform on these classes: checking membership.

There might be some inherent structure to the data we work with, which could be properties such as lexicographical orders on names—it enables us to sort student names, for example. Other structure we add on top of this. We add structure by defining classes as subsets of student names. There is even a third level of structure: how we represent the classes on our computer.

The first level of structure—inherent in the data we work with—is not something we have much control over. We might be able to exploit it in various ways, but otherwise, it is just there. When it comes to designing algorithms and data structures, this structure is often simple information; if there is order in our data, we can sort it, for example. Different algorithms and different data structures make various assumptions about the underlying data, but most general algorithms and data structures make few assumptions. When I make assumptions in this book, I will make those assumptions explicit.

The second level of structure—the structure we add on top of the universe of possible data points—is information in addition to what just exists out there in the wild; this can be something as simple as defining classes as subsets of student names. It is structure we add to data for a purpose, of course. We want to manipulate this structure and use it to answer questions while we evaluate our programs. When it comes to algorithmic theory, what we are mainly interested in at this level is which operations are possible on the data. If we represent classes as sets of student names, we are interested in testing membership to a set. To construct the classes, we might also want to be able to add elements to an existing set. That might be all we are interested in, or we might also want to be able to remove elements from a set, get the intersection or union of two sets, or do any other operation on sets.

What we can do with data in a program is largely defined by the operations we can do on structured data; how we implement the operations is less important. That might affect the efficiency of the operations and thus the program, but when it comes to what is possible to program and what is not—or what is easy to program and what is hard, at least—it is the possible operations that are important.

Because it is the operations we can do on data, and now how we represent the data—the third level of structure we have—that is most important, we distinguish between the possible operations and how they are implemented. We define *abstract data structures* by the operations we can do and call different implementations of them *concrete data structures*. Abstract data structures are defined by which operations we can do on data; concrete data structures, by how we represent the data and implement these operations.

Abstract Data Structures in R

If we define abstract data structures by the operations they provide, it is natural to represent them in R by a set of generic functions. In this book, I will use the S3 object system for this.[1]

Let's say we want a data structure that represents sets, and we need two operations on it: we want to be able to insert elements into the set, and we want to be able to check if an element is found in the set. The generic interface for such a data structure could look like this:

```
insert <- function(set, elem) UseMethod("insert")
member <- function(set, elem) UseMethod("member")
```

Using generic functions, we can replace one implementation with another with little hassle. We just need one place to specify which concrete implementation we will use for an object we will otherwise only access through the abstract interface. Each implementation we write will have one function for constructing an empty data structure. This empty structure sets the class for the concrete implementation, and from here on we can access the data structure through generic functions. We can write a simple list-based implementation of the set data structure like this:

```
empty_list_set <- function() {
  structure(c(), class = "list_set")
}

insert.list_set <- function(set, elem) {
  structure(c(elem, set), class = "list_set")
}
```

---

[1]If you are unfamiliar with generic functions and the S3 system, you can check out my book *Advanced Object-Oriented Programming in R* book (Apress, 2017), where I explain all this.

```
member.list_set <- function(set, elem) {
  elem %in% set
}
```

The empty_list_set function is how we create our first set of the concrete type. When we insert elements into a set, we also get the right type back, but we shouldn't call insert.list_set directly. We should just use insert and let the generic function mechanism pick the right implementation. If we make sure to make the only point where we refer to the concrete implementation be the creation of the empty set, then we make it easier to replace one implementation with another:

```
s <- empty_list_set()
member(s, 1)
## [1] FALSE
s <- insert(s, 1)
member(s, 1)
## [1] TRUE
```

When we implement data structures in R, there are a few rules of thumb we should follow, and some are more important than others. Using a single "empty data structure" constructor and otherwise generic interfaces is one such rule. It isn't essential, but it does make it easier to work with abstract interfaces.

More important is this rule: keep modifying and querying a data structure as separate functions. Take an operation such as popping the top element of a stack. You might think of this as a function that removes the first element of a stack and then returns the element to you. There is nothing wrong with accessing a stack this way in most languages, but in functional languages, it is much better to split this into two different operations: one for getting the top element and another for removing it from the stack.

The reason for this is simple: our functions can't have side effects. If a "pop" function takes a stack as an argument, it cannot modify this stack. It can give you the top element of the stack, and it can give you a new stack where the top element is removed, but it cannot give you the top element and then modify the stack as a side effect. Whenever we want to modify a data structure, what we have to do in a functional language, is to create a new structure instead. And we need to return this new structure to the caller. Instead of wrapping query answers *and* new (or "modified") data structures in lists so we can return multiple values, it is much easier to keep the two operations separate.

Another rule of thumb for interfaces that I will stick to in this book, with one exception, is that I will always have my functions take the data structure as the first argument. This isn't something absolutely necessary, but it fits the convention for generic functions, so it makes it easier to work with abstract interfaces, and even when a function is not abstract—when I need some helper functions—remembering that the first argument is always the data structure is easier. The one exception to this rule is the construction of linked lists, where tradition is to have a construction function, cons, that takes an element as its first argument and a list as its second argument and construct a new list where the element is put at the head of the list. This construction is too much of a tradition for me to mess with, and I won't write a generic function of it, so it doesn't come into conflict with how we handle polymorphism.

Other than that, there isn't much more language mechanics to creating abstract data structures. All operations we define on an abstract data structure have some intended semantics to them, but we cannot enforce this through the language; we just have to make sure that the operations we implement actually do what they are supposed to do.

Implementing Concrete Data Structures in R

When it comes to concrete implementations of data structures, there are a few techniques we need in order to translate the data structure designs into R code. In particular, we need to be able to represent what are essentially pointers, and we need to be able to represent empty data structures. Different programming languages will have different approaches to these two issues. Some allow the definition of recursive data types that naturally handle empty data structures and pointers, others have unique values that always represent "empty," and some have static type systems to help. We are programming in R, though, so we have to make it work here.

For efficient data structures in functional programming, we need recursive data types, which essentially boils down to representing pointers. R doesn't have pointers, so we need a workaround. That workaround is using lists to define data structures and using named elements in lists as our pointers.

Consider one of the simplest data structures known to man: the linked list. If you are not familiar with linked lists, you can read about them in the next chapter, where I consider them in some detail. In short, linked lists consist of a *head*—an element we store in the list—and a *tail*—another list, one item shorter. It is a recursive definition that we can write like this:

```
LIST = EMPTY | CONS(HEAD, LIST)
```

Here EMPTY is a special symbol representing the empty list, and CONS—a traditional name for this, from the Lisp programming language—a symbol that constructs a list from a HEAD element and a tail that is another LIST. The definition is recursive—it defines LIST in terms of a tail that is also a LIST—and this in principle allows lists to be infinitely long. In practice, a list will eventually end up at EMPTY.

9

We can construct linked lists in R using R's built-in `list` data structure. That structure is *not* a linked list; it is a fixed-size collection of elements that are possibly named. We exploit named elements to build pointers. We can implement the `CONS` construction like this:

```
linked_list_cons <- function(head, tail) {
  structure(list(head = head, tail = tail),
          class = "linked_list_set")
}
```

We just construct a `list` with two elements, `head` and `tail`. These will be references to other objects—`head` to the element we store in the list, and `tail` to the rest of the list—so we are in effect using them as pointers. We then add a class to the list to make linked lists work as an implementation of an abstract data structure.

Using classes and generic functions to implement polymorphic abstract data structures leads us to the second issue we need to deal with in R. We need to be able to represent empty lists. The natural choice for an empty list would be `NULL`, which represents "nothing" for the built-in `list` objects, but we can't get polymorphism to work with `NULL`. We can't give `NULL` a class. We could, of course, still work with `NULL` as the empty list and just have classes for non-empty lists, but this clashes with our desire to have the empty data structures being the one point where we decide concrete data structures instead of just accessing them through an abstract interface. If we didn't give empty data structures a type, we would need to use concrete update functions instead. That could make switching between different implementations cumbersome. We really *do* want to have empty data structures with classes.

The trick is to use a sentinel object to represent empty structures. *Sentinel* objects have the same structure as non-empty data structure objects—which has the added benefit of making some implementations easier to write—but they are recognized as representing "empty." We construct a sentinel as we would any other object, but we remember it for future reference. When we create an empty data structure, we always return the same sentinel object, and we have a function for checking emptiness that examines whether its input is identical to the sentinel object. For linked lists, this sentinel trick would look like this:

```
linked_list_nil <- linked_list_cons(NA, NULL)
empty_linked_list_set <- function() linked_list_nil
is_empty.linked_list_set <- function(x)
  identical(x, linked_list_nil)
```

The is_empty function is a generic function that we will use for all data structures.

The identical test isn't perfect. It will consider any list element containing NA as the last item in a list as the sentinel. Because we don't expect anyone to store NA in a linked list—it makes sense to have missing data in a lot of analysis, but rarely does it make sense to store it in data structures—it will have to do.

Using a sentinel for empty data structures can also occasionally be useful for more than dispatching on generic functions. Sometimes, we actually want to use sentinels as proper objects, because it simplifies certain functions. In those cases, we can end up with associating meta-data with "empty" sentinel objects. We will see examples of this when we implement red-black search trees. If we do this, then checking for emptiness using identical will not work. If we modify a sentinel to change meta-information, it will no longer be identical to the reference empty object. In those cases, we will use other approaches to testing for emptiness.

Asymptotic Running Time

Although the operations we define in the interface of an abstract data type determine how we can use these in our programs, the *efficiency* of our programs depends on how efficient the data structure operations are.

Because of this, we often consider the time efficiency part of the interface of a data structure—if not part of the *abstract* data structure, we very much care about it when we have to pick concrete implementations of data structures for our algorithms.

When it comes to algorithmic performance, the end goal is always to reduce *wall time*—the actual time we have to wait for a program to finish. But this depends on many factors that cannot necessarily know about when we design our algorithms. The computer the code will run on might not be available to us when we develop our software, and both its memory and CPU capabilities are likely to affect the running time significantly. The running time is also likely to depend intimately on the data we will run the algorithm on. If we want to know exactly how long it will take to analyze a particular set of data, we have to run the algorithm on this data. Once we have done this, we know exactly how long it took to analyze the data, but by then it is too late to explore different solutions to do the analysis faster.

Because we cannot practically evaluate the efficiency of our algorithms and data structures by measuring the running time on the actual data we want to analyze, we use different techniques to judge the quality of various possible solutions to our problems.

One such technique is the use of *asymptotic complexity*, also known as *big-O notation*. Simply put, we abstract away some details of the running time of different algorithms or data structure operations and classify their runtime complexity according to upper bounds known up to a constant.

First, we reduce our data to its size. We might have a set with n elements, or a string of length n. Although our data structures and algorithms might use very different actual wall time to work on different data of the same size, we care only about the number n and not the details of the data. Of course, data of the same size is not all equal, so when we reduce all our information about it to a single size, we have to be a little careful about what we mean when we talk about the algorithmic complexity of a problem. Here, we usually use one of two approaches: we speak of the *worst-case* or the *average/expected* complexity. The worst-case runtime complexity of an algorithm is the longest running time we can expect from it on any data of size n. The expected runtime complexity of an algorithm is the mean running time for data of size n, assuming some distribution over the possible data.

Second, we do not consider the *actual* running time for data of size n—where we would need to know exactly how many operations of different kinds would be executed by an algorithm, and how long each kind of operation takes to execute. We just count the number of operations and consider them equal. This gives us some function of n that tells us how many operations an algorithm or operation will execute, but not how long each operation takes. We don't care about the details when comparing most algorithms because we only care about asymptotic behavior when doing most of our algorithmic analysis.

By *asymptotic behavior*, I mean the behavior of functions when the input numbers grow large. A function $f(n)$ is an asymptotic upper bound for another function $g(n)$ if there exists some number N such that $g(n) \le f(n)$ whenever $n > N$. We write this in big-O notation as $g(n) \in O(f(n))$ or $g(n) = O(f(n))$ (the choice of notation is a little arbitrary and depends on which textbook or reference you use).

The rationale behind using asymptotic complexity is that we can use it to reason about how algorithms will perform when we give them larger data sets. If we need to process data with millions of data points, we might be about to get a feeling for their running time through experiments with tens or hundreds of data points, and we might conclude that one algorithm outperforms another in this range. But that does not necessarily reflect how the two algorithms will compare for much larger data. If one algorithm is asymptotically faster than another, it *will* eventually outperform the other—we just have to get to the point where n gets large enough.

A third abstraction we often use is to not be too concerned with getting the exact number of operations as a function of n correct. We just want an upper bound. The big-O notation allows us to say that an algorithm runs in any big-O complexity that is an upper bound for the actual runtime complexity. We want to get this upper bound as exact as we can, to properly evaluate different choices of algorithms, but if we have upper and lower bounds for various algorithms, we can still compare them. Even if the bounds are not tight, if we can see that the upper bound of one algorithm is better than the lower bound of another, we can reason about the asymptotic running time of solutions based on the two.

To see the asymptotic reasoning in action, consider the set implementation we wrote earlier:

```
empty_list_set <- function() {
  structure(c(), class = "list_set")
}

insert.list_set <- function(set, elem) {
  structure(c(elem, set), class = "list_set")
}

member.list_set <- function(set, elem) {
  elem %in% set
}
```

It represents the set as a vector, and when we add elements to the set, we simply concatenate the new element to the front of the existing set. *Vectors*, in R, are represented as contiguous memory, so when we construct new vectors this way, we need to allocate a block of memory to contain the new vector, copy the first element into the first position, and then copy the entire old vector into the remaining positions of the new vector. Inserting an element into a set of size n, with this implementation, will take time $O(n)$—we need to insert $n+1$ set elements into newly allocated blocks of memory. Growing a set from size 0 to size n by repeatedly inserting elements will take time $O(n^2)$.

The membership test, elem %in% set, runs through the vector until it either sees the value elem or reaches the end of the vector. The best case would be to see elem at the beginning of the vector, but if we consider worst-case complexity, this is another $O(n)$ runtime operation.

As an alternative implementation, consider linked lists. We insert elements in the list using the cons operation, and we check membership by comparing elem with the head of the list. If the two are equal, the set contains the element. If not, we check whether elem is found in the rest of the list. In a pure functional language, we would use recursion for this search, but here I have just implemented it using a while loop:

```
insert.linked_list_set <- function(set, elem) {
  linked_list_cons(elem, set)
}

member.linked_list_set <- function(set, elem) {
  while (!is_empty(set)) {
    if (set$head == elem) return(TRUE)
    set <- set$tail
  }
  return(FALSE)
}
```

The insert operation in this implementation takes constant time. We create a new list node and set the head and tail in it, but unlike the vector implementation, we do not copy anything. For the linked list, inserting elements is an $O(1)$ operation. The membership check, though, still runs in $O(n)$ because we still do a linear search.

Experimental Evaluation of Algorithms

Analyzing the asymptotic performance of algorithms and data structures is the only practical approach to designing programs that work on very large data, but it cannot stand alone when it comes to writing efficient code. Some experimental validation is also needed. We should always perform experiments with implementations to 1) be informed about the performance constants hidden beneath the big-O notation, and 2) to validate that the performance is as we expect it to be.

For the first point, remember that just because two algorithms are in the same big-O category—say, both are in $O(n^2)$—that doesn't mean they have the same wall-time performance. It means that both algorithms are asymptotically bounded by some function $c \cdot n^2$ where c is a constant. Even if both are running in quadratic time, so that the upper bound is actually tight, they could be bounded by functions with very different constants. They may have the same asymptotic complexity, but in practice, one could be much faster than the other. By experimenting with the algorithms, we can get a feeling, at least, for how the algorithms perform in practice.

Experimentation also helps us when we have analyzed the *worst case* asymptotic performance of algorithms, but where the data we actually want to process is different from the worst possible data. If we can create samples of data that resemble the actual data we want to analyze, we can get a feeling for how close it is to the worst case, and perhaps find that an algorithm with worse *worst case* performance actually has better *average case* performance.

As for point number two for why we want to experiment with algorithms, it is very easy to write code with a different runtime complexity than we expected, either because of simple bugs or because we are programming in R, a very high-level language, where language constructions potentially hide complex operations. Assigning to a vector, for example, is not a simple constant time operation if more than one variable refers to the vector. Assignment to vector elements potentially involves copying the entire vector. Sometimes it is a constant time operation; sometimes it is a linear time operation. We can deduce what it will be by carefully reading the code, but it is human to err, so it makes sense always to validate that we have the expected complexity by running experiments.

In this book, I will use the `microbenchmark` package to run performance experiments. This package lets us run a number of executions of the same operation and get the time it takes back in nanoseconds. I don't need that fine a resolution, but it is nice to be able to get a list of time measurements. I collect the results in a `tibble` data frame from which I can summarize the results and plot them later. The code I use for my experiments is as follows:

```r
library(tibble)
library(microbenchmark)

get_performance_n <- function(
  algo
, n
, setup
, evaluate
, times
, ...) {

  config <- setup(n)
  benchmarks <- microbenchmark(evaluate(n, config),
                               times = times)
  tibble(algo = algo, n = n,
         time = benchmarks$time / 1e9) # time in sec
}

get_performance <- function(
  algo
, ns
, setup
, evaluate
, times = 10
, ...) {
```

```
  f <- function(n)
    get_performance_n(algo, n, setup, evaluate,
                       times = times, ...)
  results <- Map(f, ns)
  do.call('rbind', results)
}
```

The performance experiment functions let me specify a function for setting up an experiment and another for running the experiment. If I want to evaluate the time it takes to construct a set of the numbers from one up to n, I can use the setup function to choose the implementation—based on their respective empty structures—and I can construct the sets in the evaluate function:

```
setup <- function(empty) function(n) empty
evaluate <- function(n, empty) {
  set <- empty
  elements <- sample(1:n)
  for (elm in elements) {
    set <- insert(set, elm)
  }
}

ns <- seq(1000, 5000, by = 500)
performance <- rbind(
  get_performance("list()", ns,
                  setup(empty_list_set()), evaluate),
  get_performance("linked list", ns,
                  setup(empty_linked_list_set()), evaluate)
)
```

I permute the elements I insert in the sets to avoid any systematic bias in how the data is added to the sets. There isn't any with the two implementations we have here, but for many data structures there are, so this is a way of getting an average case complexity instead of a best-case or worst-case performance.

Running the performance measuring code with these two functions and the two set implementations, I get the results I have plotted in Figure 2-1:

```
library(ggplot2)
ggplot(performance, aes(x = n, y = time, colour = algo)) +
  geom_jitter() +
  geom_smooth(method = "loess",
                span = 2, se = FALSE) +
  scale_colour_grey("Data structure", end = 0.5) +
  xlab(quote(n)) + ylab("Time (sec)") + theme_minimal()
```

In this figure, we can see what we expected from the asymptotic runtime analysis. The two approaches are not that different for small sets, but as the size of the data grows, the list implementation takes relatively longer to construct a set than the linked list implementation.

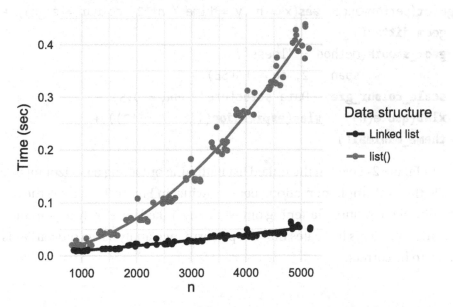

Figure 2-1. *Direct comparison of the two set construction implementations*

We cannot directly see from Figure 2-1 that one data structure takes linear time and the other quadratic time. That can be hard to glean just from a time plot. To make it easier to see, we can divide the actual running time by the expected asymptotic running time. If we have the right asymptotic running time, the time usage divided by expected time should flatten out around the constant that the asymptotic function is multiplied with. So, if the actual running time is $c \cdot n^2$, then dividing the running time by n^2 we should see the plot flatten out around $y = c$.

In Figure 2-2 we see the time divided by the size of the set, and in Figure 2-3 the time divided by the square of the size of the set:

```
ggplot(performance, aes(x = n, y = time / n, colour = algo)) +
  geom_jitter() +
  geom_smooth(method = "loess",
              span = 2, se = FALSE) +
  scale_colour_grey("Data structure", end = 0.5) +
  xlab(quote(n)) + ylab("Time / n") + theme_minimal()

ggplot(performance, aes(x = n, y = time / n**2, colour = algo)) +
  geom_jitter() +
  geom_smooth(method = "loess",
              span = 2, se = FALSE) +
  scale_colour_grey("Data structure", end = 0.5) +
  xlab(quote(n)) + ylab(expression(Time / n**2)) +
  theme_minimal()
```

In Figure 2-2 we see the linked list flattening out along a horizontal line while the list implementation keeps growing; in Figure 2-3 we see the list implementation flattening out while the linked list slowly asymptotes towards zero. This indicates that the performance analysis we did earlier is likely to be correct.

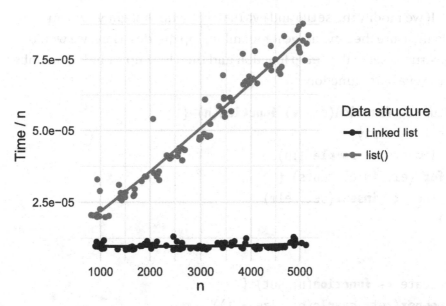

Figure 2-2. *The two set construction implementations with time divided by input size*

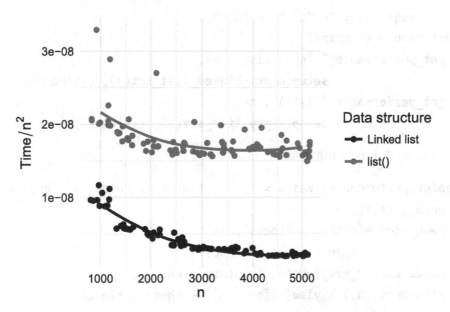

Figure 2-3. *The two set construction implementations with time divided by input size squared*

If we modify the `setup` and `evaluate` functions slightly, we can also measure the time usage for membership queries. Here, we would construct a set in the `setup` function and then look up a random member in the `evaluate` function:

```
setup <- function(empty) function(n) {
  set <- empty
  elements <- sample(1:n)
  for (elm in elements) {
    set <- insert(set, elm)
  }
  set
}
evaluate <- function(n, set) {
  member(set, sample(n, size = 1))
}

ns <- seq(10000, 50000, by = 10000)
performance <- rbind(
  get_performance("linked list", ns,
                  setup(empty_linked_list_set()), evaluate),
  get_performance("list()", ns,
                  setup(empty_list_set()), evaluate))
```

Figure 2-4 plots the results:

```
ggplot(performance, aes(x = n, y = time / n, colour = algo)) +
  geom_jitter() +
  geom_smooth(method = "loess",
              span = 2, se = FALSE) +
  scale_colour_grey("Data structure", end = 0.5) +
  xlab(quote(n)) + ylab("Time / n") + theme_minimal()
```

I have plotted the time usage divided by *n* because we expect both implementations to have linear time member queries. This is also what we see, but we also see that the linked list is slower and has a much larger variance in its performance. Although both data structures have linear time member queries, the `list` implementation is faster in practice. For member queries, as we have seen, it is certainly not faster when it comes to constructing sets one element at a time.

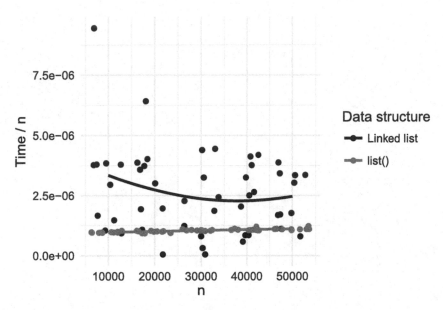

Figure 2-4. *Comparison of member queries for the two set implementations; time divided by input size*

CHAPTER 3

Immutable and Persistent Data

What prevents us from implementing traditional imperative-language data structures in R is the immutability of data. As a general rule, you can modify environments—so you can assign to variables—but you cannot modify actual data. Whenever R makes it look like you are changing data, it is lying. When you assign to an element in a vector

```
x[i] <- v
```

the vector will look modified to you, but behind the curtain, R has really replaced the vector that x refers to with a new copy, identical to the old x except for element number i. It tries to do this efficiently, so it will only copy the vector if there are other references to it, but conceptually, it still makes a copy.

Now, you could reasonably argue that there is little difference between actually modifying data and simply having the illusion of changing data, and you would be right—except that the illusion is only skin deep. Because R creates the illusion by making copies of data and assigning the copies to variables in the local environment, it doesn't affect other references to the original data. Data you pass to a function as a parameter will be referenced by a local function variable. If we "modify" such data, we are changing the

© Thomas Mailund 2017
T. Mailund, *Functional Data Structures in R*, https://doi.org/10.1007/978-1-4842-3144-9_3

local environment—the caller of the function has a different reference to the same data, and that reference is to the original data that will not be affected by what we do with the local function environment in any way. R is not entirely side-effect free, as a programming language, but side effects are contained to I/O, random number generation, and affecting variable-value bindings in environments. Modifying actual data is not something we can do via function side effects.[1] If we want to update a data structure, we have to do what R does when we try to modify data: we need to build a *new* data structure, looking like the one we wanted to change the old one into. Functions that should update data structures need to construct new versions and return them to the caller.

Persistent Data Structures

When we update an imperative data structure we typically accept that the old version of the data structure will no longer be available, but when we update a functional data structure, we expect that both the old and new versions of the data structure will be available for further processing. A data structure that supports multiple versions is called *persistent*, whereas a data structure that allows only a single version at a time is called *ephemeral*. What we get out of immutable data is persistent data structures; these are the natural data structures in R.

---

[1]Strictly speaking, we *can* create side effects that affect data structures—we just have to modify environments. The reference class system, R6, emulates objects with a mutable state by updating environments, and we can do the same via closures. When we get to Chapter 4, where we will implement queues, I'll introduce side effects of member queries, and there we will use this trick. Unless we represent all data structures by collections of environments, though, the method only gets us so far. We still need to build data structures without modifying data—we just get to remember the result in an environment we constructed for this purpose.

Not all types of data structures have persistent versions of themselves, and some persistent data structures can be less efficient than their ephemeral counterparts, but constructing persistent data structures is an active area of research, so there are data structures enough to pick from when you need one. When there are no good choices of persistent data structures, though, it *is* possible to implement ephemeral structures, if we exploit environments, which *are* mutable.

To see what I mean by data structures being persistent in R, let's look at the simple linked list again. I've defined it as follows, using slightly shorter names than earlier, now that we don't need to remind ourselves that it is a linked list, and I'm using the sentinel trick to create the "empty" list:

```
is_empty <- function(x) UseMethod("is_empty")

list_cons <- function(elem, lst)
  structure(list(item = elem, tail = lst), class = "linked_list")

list_nil <- list_cons(NA, NULL)
is_empty.linked_list <- function(x) identical(x, list_nil)
empty_list <- function() list_nil

list_head <- function(lst) lst$item
list_tail <- function(lst) lst$tail
```

With these definitions, we can create three lists like this:

```
x <- list_cons(2, list_cond(1, empty_list()))
y <- list_cons(3, x)
z <- list_cons(4, empty_list())
```

The lists will be represented in memory as shown in Figure 3-1. In the figure, I have shown the content of the lists, the head of each, in the white boxes, and the tail pointer as a grey box and an arrow. I have explicitly shown the empty list sentinel in this figure, but in future figures, I will leave it out. The variables, x, y, and z are shown as pointers to the lists.

For x and z, the lists were created by updating the empty list; for y, the list was created by updating x. But as we can clearly see, the updated lists are still there. We just need to keep a pointer to them to get them back. That is the essence of persistent data structures and how we need to work with data structures in R.

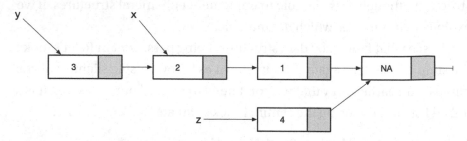

Figure 3-1. *Memory layout of linked lists*

List Functions

When we implemented a set through linked lists we saw how to add and search in a list. To get a feeling of working with immutable data structures we will try to implement a few more functions manipulating lists. We can start simply by writing a function for reversing a list. This is slightly more complicated than just searching in a list because we will have to construct the reversed list from the wrong end, so to speak. We need to first construct the list that contains the last element of the input list and the empty list. Then we need to put the second last elements at the head of this list, and so on.

When writing a function that operates on persistent data, I always find it easiest to think in terms of recursion. It may not be immediately obvious how to reverse a list as a recursive function, though. If we recurse all the way down to the end of the list, we get hold of the first element we should have in the reversed list, but how do we then fit that into the list we construct going up in the recursion again? There is no simple way to do this. We can, however, use the trick of bringing an accumulator with us in the recursive calls and construct the reversed list using that. If you are not familiar with accumulators in recursive functions, I cover it in some detail in my book *Advanced Object-Oriented Programming in R* (Apress, 2017), but you can probably follow the idea in the following code. The idea is that the variable acc contains the reversed list we have constructed so far. When we get to the end of the recursion, we have the entire reversed list in acc so we can just return it. Otherwise, we can recurse on the remaining list but put the head element at the top of the accumulator. With a recursive helper function, the list reversal can look like this:

```r
list_reverse_helper <- function(lst, acc) {
  if (is_empty(lst)) acc
  else list_reverse_helper(list_tail(lst),
                           list_cons(list_head(lst), acc))
}
list_reverse_rec <- function(lst)
  list_reverse_helper(lst, empty_list())
```

The running time of this function is linear—we need to run through the entire original list, but each operation we do when we construct the new list takes constant time.

I have shown the iterations for reversing a list of length three in Figure 3-2. In this figure, I have not shown the empty sentinel string—I just show the empty string as a pointer to nothing. But you will see how the variable lst refers to different positions in the original list as we recurse, whereas the original list does not change at all, as we build a new list pointed to by acc.

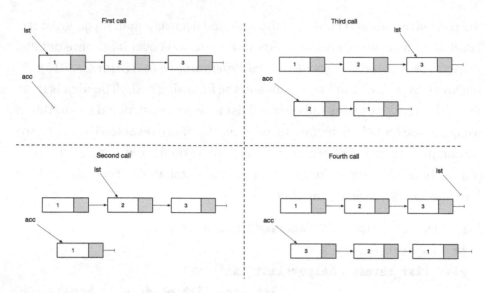

Figure 3-2. *Iterations in the recursive list reversal*

In a purely functional programming language, this would probably be the best approach to reversing a list. The function uses tail recursion (again, you can read about that in my other book), so it is essentially a loop we have written. Unfortunately, R does *not* implement tail recursion, so we have a potential problem. If we have a very long list, we can run out of stack space before we finish reversing it. We can, however, almost automatically translate tail recursive functions into loops, and a loop version for reversing a list would then look like this:

```
list_reverse_loop <- function(lst) {
  acc <- empty_list()
  while (!is_empty(lst)) {
    acc <- list_cons(list_head(lst), acc)
    lst <- list_tail(lst)
  }
  acc
}
```

We can perform some experiments to explore the performance of the two solutions. With the performance measurement functions described in chapter 2, we can set up the experiments like this:

```
setup <- function(n) {
  lst <- empty_list()
  elements <- sample(1:n)
  for (elm in elements) {
    lst <- list_cons(elm, lst)
  }
  lst
}
evaluate_rec <- function(n, lst) {
  list_reverse_rec(lst)
}
evaluate_loop <- function(n, lst) {
  list_reverse_loop(lst)
}
```

We construct lists of the two types in the setup function and then reverse them in the evaluate function. We run the actual experiments like this:

```
ns <- seq(100, 500, by = 50)
performance <- rbind(
  get_performance("recursive", ns, setup, evaluate_rec,
  times = 25),
  get_performance("loop", ns, setup, evaluate_loop, times = 25)
)
```

We can then plot the results using ggplot2 like this:

```
library(ggplot2)
ggplot(performance, aes(x = n, y = time / n, colour = algo)) +
  geom_jitter() +
  geom_smooth(method = "loess", se = FALSE, span = 2) +
  scale_colour_grey("Data structure", end = 0.5) +
  xlab(quote(n)) + ylab("Time / n") + theme_minimal()
```

The results are shown in Figure 3-3.

If the lists are short, there is no immediate benefit in using one solution over the other. There is some overhead in function calls, but there is also some overhead in loops, and the two solutions work equally well for short lists. Whenever I can get away with it, I prefer recursive solutions—I find them easier to implement and simpler to understand—but the loop version will be able to deal with much longer lists than the recursive one.

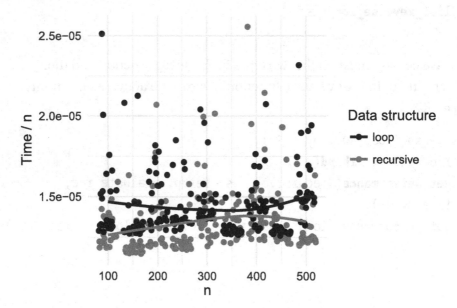

Figure 3-3. *Looping versus recursive reversal of lists*

Another thing we might want to do with lists is concatenate two of them. With mutable pointers, this is something we can do in constant time if we have pointers to both the beginning and end of our linked lists. With immutable data it becomes a linear time function—we need to construct the new concatenated list without modifying the two original lists, so we need to move to the end of the first list and then create a new one that contains the same elements followed by the second list. Strictly speaking, it is not a linear time algorithm in the length of both lists, but it is linear in the length of the first list.

Again, it is easiest to construct the function recursively. Here, the base case is when the first list is empty. Then, the concatenation of the two lists is just the second list. Otherwise, we have to put the head of the first list in front of the concatenation of the tail of the first list and the entire second list. As an R function, we can implement that idea like this:

```r
list_concatenate <- function(l1, l2) {
  if (is_empty(l1)) l2
  else list_cons(list_head(l1),
                 list_concatenate(list_tail(l1), l2))
}
```

The new list that we construct contains l2 as the last part of it. We do not need to copy this—it is, after all, an immutable data structure, so there is no chance of it changing in the future—but we are putting a new copy of l1 in front of it. The structure of the lists after we concatenate them is shown in Figure 3-4. The two original lists are alive and well, and we have a new concatenated version.

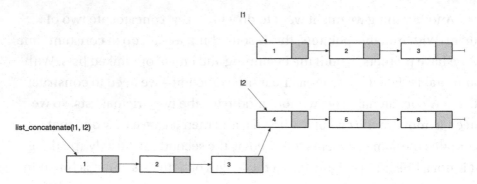

Figure 3-4. *Persistent lists in list concatenation*

It is a little harder to implement concatenation without recursion.
We construct the new list as we return from the recursive calls, so if we
want to implement this iteratively, we need to emulate the call stack. We
can do this, however, following the way we implemented list reversal: we
can construct a reversal of the first list moving down the list, and once we
read the end of the first list we can build the result of the concatenation by
putting head elements in front of the new list. We can implement a looping
version this way:

```
list_concatenate_loop <- function(l1, l2) {
  rev_l1 <- empty_list()
  while (!is_empty(l1)) {
    rev_l1 <- list_cons(list_head(l1), rev_l1)
    l1 <- list_tail(l1)
  }
  result <- l2
  while (!is_empty(rev_l1)) {
    result <- list_cons(list_head(rev_l1), result)
    rev_l1 <- list_tail(rev_l1)
  }
  result
}
```

To experiment with concatenation, we can use the same `setup` function as before, but change the `evaluate` functions to these:

```
evaluate_rec <- function(n, lst) {
  list_concatenate(lst, lst)
}
evaluate_loop <- function(n, lst) {
  list_concatenate_loop(lst, lst)
}
```

The actual performance runs and plotting code is similar to before.

The loop version first has to construct the reversed of the first list and then the concatenated list, so it does more work than the recursive version and therefore is slower (Figure 3-5). I think you would also agree that the loop version is quite a bit more complex than the recursive one. It can, however, deal with longer lists because it doesn't require a deep call stack. Still, unless we run out of stack space in the recursive function, the recursive version is probably the better choice.

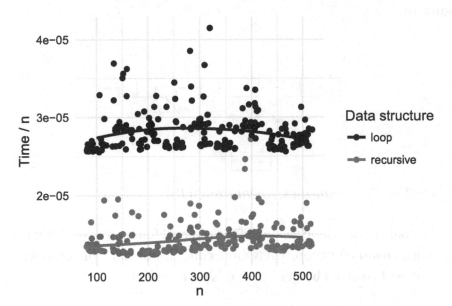

Figure 3-5. *Time comparison of the two concatenation functions*

What about removing elements from a list? Again, we cannot modify lists, so "removing" means constructing a new list, and once again, the easiest approach to solving the problem is to write a recursive function. The base case is removing an element from an empty list. That is easily done because we can just return the empty list. Otherwise, we have two cases—assuming we only want to remove the first occurrence of an element, as I will assume there. If the head of the list is equal to the element we want to remove, we just return the tail of the list. Otherwise, we need to concatenate the current head to a recursive call to delete the element from the tail of the list. The solution could look like this:

```
list_remove <- function(lst, elm) {
  if (is_empty(lst)) lst
  else if (list_head(lst) == elm) list_tail(lst)
  else list_cons(list_head(lst), list_remove(list_tail(lst), elm))
}
```

An example of the new list constructed by a removal is shown in Figure 3-6.

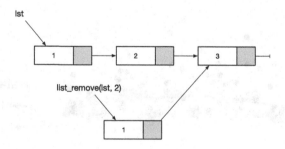

Figure 3-6. *Removing an element from a list*

We could make a loop version of the removal function as well, but like the concatenation function, it gets more complicated than the recursive solution, and I will not bother with it this time around.

The general problem with both the concatenation and removal functions that prevent us from writing simple loop versions is that they are not tail recursive. It is generally easy to translate tail recursive functions into loops, but in both the concatenation and the removal functions, we need to build a new list as part of the recursion, and that complicates matters. There are general approaches to get around this and translate your functions into tail recursive ones, using what are called *continuations*, but this adds an overhead in several ways, not least because you would still have to emulate actual tail recursion to avoid reaching the stack limit. So *if* recursive functions cause you problems, you are better off trying to write a custom looping version instead. That being said, because list functions typically have linear running time, if you have to worry about stack limits, you should probably reconsider the data structure in any case and use a more efficient one. For functional data structures, that almost always means using a tree instead of a list.

Trees

Linked lists are the workhorses of much functional programming, but when it comes to efficiency, trees are usually better alternatives. In fact, most of the remaining chapters of this book will use variations of trees to implement various abstract data structures. For now, though, we will just consider how to construct and traverse trees. For this, we implement a simple, unbalanced, search tree. Search trees are trees with the following properties:

1. Each node in the tree contains a number (or any other element from an ordered set).

2. Each node has two children—we call them the *left* and *right* children. These children are subtrees.

3. All elements in the left subtree of a node contain
    values smaller than the element in the node.

4. All elements in the right subtree of a node contain
    values larger than the element in the node.

An example of a search tree, containing the elements 1, 3, 4, 6, and 9, is
shown in Figure 3-7.

The asymptotic efficiency of search trees comes from how they are
balanced. Searching in a search tree involves recursing down the tree
structure and will take time proportional to the depth we search. The worst
case for a tree would be one that is essentially a list—we could get such a
tree if all nodes only had a single child—where the search time would be
linear, just like searching in lists. We can typically balance them, though,
in which case the depth is only going to be logarithmic in the number of
elements we store in them. To see this, consider a binary tree where at
each level all nodes have two children. That would double the number
of elements we have in the tree at each level, and we can only double a
number $\log(n)$ many times before we have n elements.

The tree we construct in this chapter will not necessarily be balanced.
We implement none of the tricks needed to keep it balanced, so the depth
of the tree will depend on the order in which we insert elements.

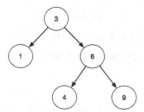

Figure 3-7. *An example of a search tree*

Anyway, we construct our tree out of nodes that contain a value and a
reference to the left and to the right subtrees. As usual, we create a sentinel

object to represent an empty tree. So our tree implementation could look like this:

```
search_tree_node <- function(
  value
  , left = empty_search_tree()
  , right = empty_search_tree()
) {
  structure(list(left = left, value = value, right = right),
            class = c("unbalanced_search_tree"))
}

empty_search_tree <- function()
  search_tree_node(NA, NULL, NULL)
is_empty.unbalanced_search_tree <- function(x)
  is.null(x$left) && is.null(x$right)
```

For the empty tree sentinel, we do not use identical for the is_empty function. This is because we want to be able to modify empty trees and still have them be empty. We will see a use for this shortly.

We want three generic functions for working with sets in general, so we define functions for inserting and removing elements in a set and a function for testing membership in a set. All three will be implemented for our search tree with complexity proportional to the depth of the tree:

```
insert <- function(x, elm) UseMethod("insert")
remove <- function(x, elm) UseMethod("remove")
member <- function(x, elm) UseMethod("member")
```

Of these three functions, the member function is the simplest to implement for a search tree. The invariant for search trees tells us that if we have a node and the element we are looking for is smaller than the value there, then the element must be found in the left search tree if it exists; otherwise, it must be found in the right search tree. This leads naturally to a recursive function. The base case is when we search in an empty tree: we will always answer that the element is not found there. Otherwise, we essentially have to check three possibilities: we might have found the node where the element is, or we have to search to the left, or we have to search to the right. We can implement that function thus:

```
member.unbalanced_search_tree <- function(x, elm) {
  if (is_empty(x)) return(FALSE)
  if (x$value == elm) return(TRUE)
  if (elm < x$value) member(x$left, elm)
  else member(x$right, elm)
}
```

Calling recursively on a generic function is slightly inefficient, though. It requires that we do the dynamic dispatch based on the class of the tree parameter in each recursive call, even though we know that the type is a search tree. We can fix this by splitting the recursive function and the member function like this:

```
st_member <- function(x, elm) {
  if (is_empty(x)) return(FALSE)
  if (x$value == elm) return(TRUE)
  if (elm < x$value) st_member(x$left, elm)
  else st_member(x$right, elm)
}
member.unbalanced_search_tree <- function(x, elm) {
  st_member(x, elm)
}
```

We can actually do even better. In this solution, we do comparisons with each recursive call if we are not at an empty tree: we compare for equality with the element we are looking for, and we check whether it is less than the value in the node. We can delay one of the comparisons and halve the number of comparisons. We just have to remember the last element that *could* be the value that we are looking for. We can then call recursively to the left or right after just checking whether the element in the node is larger than the element we are searching for. If we get all the way down to an empty tree, we check whether the element is the one we could have checked equality on earlier. That solution looks like this:

```
st_member <- function(x, elm, candidate = NA) {
  if (is_empty(x)) return(!is.na(candidate) && elm == candidate)
  if (elm < x$value) st_member(x$left, elm, candidate)
  else st_member(x$right, elm, x$value)
}
member.unbalanced_search_tree <- function(x, elm) {
  st_member(x, elm)
}
```

This solution reduces the number of comparisons per inner node in the tree, but it also risks going deeper in the tree than necessary because it will move past the element we are searching for.

We can evaluate the actual performance of the three solutions using this code:

```
member.unbalanced_search_tree <- function(x, elm) {
  if (is_empty(x)) return(FALSE)
  if (x$value == elm) return(TRUE)
  if (elm < x$value) member(x$left, elm)
  else member(x$right, elm)
}

st_member_slow <- function(x, elm) {
  if (is_empty(x)) return(FALSE)
  if (x$value == elm) return(TRUE)
  if (elm < x$value) st_member_slow(x$left, elm)
  else st_member_slow(x$right, elm)
}
```

41

```r
st_member_fast <- function(x, elm, candidate = NA) {
  if (is_empty(x)) return(!is.na(candidate) && elm == candidate)
  if (elm < x$value) st_member_fast(x$left, elm, candidate)
  else st_member_fast(x$right, elm, x$value)
}

setup_for_member <- function(n) {
  tree <- empty_search_tree()
  elements <- sample(1:n)
  for (elm in elements) {
    tree <- insert(tree, elm)
  }
  tree
}
evaluate_member <- function(member_func) function(n, tree) {
  elements <- sample(1:n, size = 100, replace = TRUE)
  for (elm in elements) {
    member_func(tree, elm)
  }
}
```

We construct trees with random input to get them roughly balanced. The results are shown in Figure 3-8. I have divided the running time with $\log(n)$ because that is the expected depth of a random tree and thus the expected running time for the member functions—which we see matches the actual running time as the time lines are horizontal. We clearly see that the overhead in using the generic function makes that solution slower, whereas the two other solutions have roughly the same running time in practice, with perhaps a slightly better running time for the optimised version. Because that version seems to be the best—at least it is never the worst—we should use that in our search tree implementations.

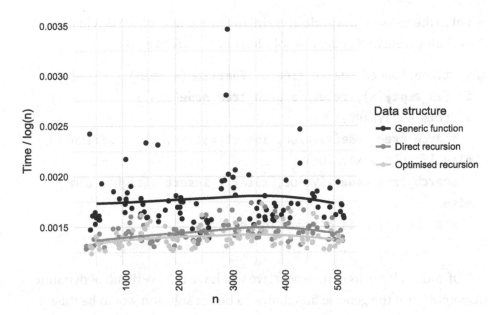

Figure 3-8. *Search tree member functions comparison*

To insert an element in a search tree, we have to, as always, construct a new structure that represents the updated tree. We can do this by searching recursively down the tree to the position where we should insert the element—going left if we are inserting an element smaller than the value in a node and going right if we are inserting a larger value—and then constructing the new tree going up the recursion. A recursive solution will work like this: if we reach an empty tree, we have found the place where we should insert the element, so we construct a new leaf and return that from the recursion. Otherwise, we check the value in the inner node. If the value is larger than the element we are inserting, we need to construct the new tree consisting of the current value, the existing right tree, and a version of the left tree where we have inserted the element. If the element in the node is smaller than the element we are inserting, we do the symmetric constructing where we insert the element in the right tree. If the element is

43

equal to the value in the node, it is already in the tree, so we can just return the existing tree. In R code, this solution could look like this:

```r
insert.unbalanced_search_tree <- function(x, elm) {
  if (is_empty(x)) return(search_tree_node(elm))
  if (elm < x$value)
    search_tree_node(x$value, insert(x$left, elm), x$right)
  else if (elm > x$value)
    search_tree_node(x$value, x$left, insert(x$right, elm))
  else
    x # the value is already in the tree
}
```

Of course, here, as with member, we will have the overhead of dynamic dispatching on the generic function, so a better solution would be this:

```r
st_insert <- function(tree, elm) {
  if (is_empty(tree))
    return(search_tree_node(elm))
  if (elm < tree$value)
    search_tree_node(tree$value,
                     st_insert(tree$left, elm),
                     tree$right)
  else if (elm > tree$value)
    search_tree_node(tree$value,
                     tree$left,
                     st_insert(tree$right, elm))
  else
    tree
}
insert.unbalanced_search_tree <- function(x, elm, ...) {
  st_insert(x, elm)
}
```

The insert function searches down the recursion for the place to put the new value and then constructs a new tree when returning from the recursion again. All the operations we do going down and up the recursion are constant time, and we only copy parts of the original tree that are directly on the search part. We just keep the existing references to the other parts of the tree. An example of an updated tree, after inserting 5 in the tree from Figure 3-7, is shown in Figure 3-9.

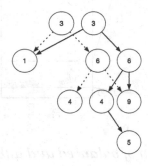

Figure 3-9. *Result of inserting 5 in the search tree from Figure 3-7*

If we have a set of elements we want to construct a search tree on, and we insert them in a random order, each element is equally likely to be put to the left or right of the root, on average, so we end up with a fairly balanced tree. This is why we got flat plots when we divided by the logarithm of the tree size as we compared membership test functions. However, if we were inserting the element in increasing order, we would always be inserting the next element in the rightmost child of the tree, and we would end up with a tree with linear depth.

If we construct search trees by inserting one element at a time, the running time would be the number of elements multiplied by the average depth. If the tree is unbalanced, this becomes an $O(n^2)$ constructing time because the average tree depth will be $n/2$. The contrast between inserting the elements in a random order or in increasing order is shown in Figure 3-10.

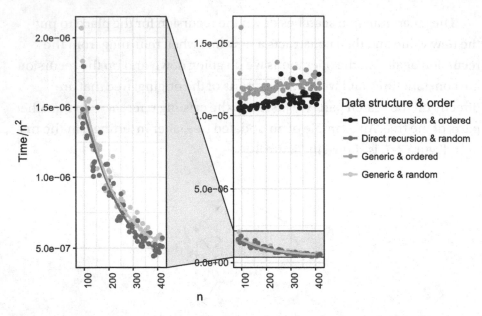

Figure 3-10. *Comparison of balanced and unbalanced search tree constructions*

For these experiments I used the following `setup` and `evaluate` functions:

```
setup_for_construction <- function(n) n
evaluate_construction <- function(order, ins) function(n, x) {
  tree <- empty_search_tree()
  elements <- order(1:n)
  for (elm in elements) {
    tree <- ins(tree, elm)
  }
  tree
}
```

The `setup` function doesn't do any real setup, it just returns the size to return *something*, but it isn't used by the `evaluate` function. The `evaluate` function is a higher order function that takes two functions—one for specifying the order for the data to insert into the search tree and one for doing the insertion—and returns the function used for the actual runtime evaluation. I use the functions as shown below, where I use the higher-order `evaluate_construction` to vary both the insertion order and the insert function. Here, I used the `insert` generic function that called itself recursively and the `st_insert` function shown earlier:

```
ns <- seq(100, 400, by = 50)
construction_performance <- rbind(
  get_performance("Generic & ordered", ns,
                  setup_for_construction,
                  evaluate_construction(identity, insert)),
  get_performance("Direct recursion & ordered", ns,
                  setup_for_construction,
                  evaluate_construction(identity, st_insert)),
  get_performance("Generic & random", ns,
                  setup_for_construction,
                  evaluate_construction(sample, insert)),
  get_performance("Direct recursion & random", ns,
                  setup_for_construction,
                  evaluate_construction(sample, st_insert))
)
```

To plot the results, I used the `facet_zoom` function from the `ggforce` package to zoom in on the experiments with random order insertion. These experiments would not be readable if we didn't zoom in on them:

```
library(ggforce)
construction_performance %>%
  mutate(is_random = grepl(".*random", algo)) %>%
  ggplot(aes(x = n, y = time / n^2, colour = algo)) +
  geom_jitter() +
  geom_smooth(method = "loess", se = FALSE, span = 1) +
```

```
scale_colour_grey("Data structure & order", end = 0.8) +
xlab(quote(n)) + ylab(expression(Time / n**2)) +
facet_zoom(y = is_random, zoom.size = 1) +
theme_bw() +
theme(axis.text.x = element_text(angle = 90, hjust = 1))
```

Again, we see the overhead involved in recursing on the generic function—so we should use the non-recursive generic function that calls st_insert—and we see that inserting elements in order is *much* slower than inserting them in random order. It is actually slightly worse than the running time suggests: because the unbalanced trees can be linear in depth, the recursive functions we use to manipulate them can run into problems with stack depths long before their balanced counterparts. This is why I have only run experiments up to size 450—at 500 I run out of stack space. We return to techniques for keeping trees balanced in sec. 6. For now, we just continue with the simple operations, and we have made it to the last function we need to implement: removal of elements.

Removing elements is the most complex function we need to implement for our search trees. The basic pattern is the same as for insertion: we search for the element to remove going down a recursion and then construct an updated tree going up the recursion again. The complication comes when we actually have to remove an element. Removing leaves is easy: we can just return an empty tree from the recursion when we construct the updated tree. It is also easy to remove nodes with a single child: here we can just return the child, and it will be inserted in the correct position in the tree we are constructing. When the element we need to remove is in an inner node with two children, though, we cannot easily update the tree to an inner node with references to the two children but not the value we want to remove.

What we want to do is to replace the value we want to delete with another value already found in the tree—a value greater than all in the left subtree and smaller than all in the right subtree. The trick is to get hold of the leftmost node in the right subtree. This node contains the smallest value in the right subtree, so if we put this value in the new inner node, and remove it from the right subtree, then the invariant will be satisfied. The leftmost node in the subtree is larger than all values in the left subtree—because it is currently in the right subtree—and it will be smaller than all the other values in the right subtree because it is the leftmost value. Also, it is easy to remove the leftmost node in a tree because it can have at most one (right) child, and those cases we can handle easily.

We need a helper function to find the leftmost node in a subtree, but after that, we just have to handle different cases when we delete an element. If it isn't in the tree, we eventually hit an empty tree in the search, and we just return that. Otherwise, if we find the element, we handle the cases where it has at most one child directly, and we do the "leftmost trick" otherwise. The remaining cases are just handling when we are still searching down in the recursion: we need to create a node that contains the original left or right subtree—depending on the value in the node—and use the result of removing the element in the other subtree. The full implementation can look like this:

```
st_leftmost <- function(x) {
  while (!is_empty(x)) {
    value <- x$value
    tree <- x$left
  }
  value
}

st_remove <- function(tree, elm) {
  # if we reach an empty tree, there is nothing to do
  if (is_empty(tree)) return(tree)
```

```
  if (tree$value == elm) {
    a <- tree$left
    b <- tree$right
    if (is_empty(a)) return(b)
    if (is_empty(b)) return(a)

    s <- st_leftmost(tree$right)
    return(search_tree_node(s, a, st_remove(b, s)))
  }

  # we need to search further down to remove the element
  if (elm < tree$value)
    search_tree_node(tree$value, st_remove(tree$left, elm),
    tree$right)
  else # (elm > tree$value)
    search_tree_node(tree$value, tree$left,
    st_remove(tree$right, elm))
}

remove.unbalanced_search_tree <- function(x, elm, ...) {
  st_remove(x, elm)
}
```

This time around, I didn't bother with writing a version that uses the generic function for recursion—by now we know that this will be less efficient than calling a plain function from the generic function and having the plain function handle the recursions.

Figure 3-11 shows the result of removing 6 from the tree in Figure 3-7. When the search finds the node that contains 6, it will go down and get the leftmost node in the right subtree, which is the leaf containing 9. It then deletes 9 from the right subtree—the result will be an empty tree— and constructs a new node that contains the value 9, has the subtree containing 4 as its left subtree, and has the empty subtree as the right subtree. It then creates a copy of the nodes above the removal point going up the recursion to construct the final updated tree.

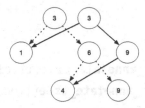

Figure 3-11. *Removing an element from a subtree*

For debugging purposes, or just to understand how a data structure is modified by various operations, it can be helpful to print or plot it. For simple lists that represent sequential data, printing the elements they contain is the simplest, but for trees and more complex data structures, we need to plot.

A powerful tool for manipulating graphs is the `tidygraph` package, and a powerful tool for plotting them is the `ggraph` extension to `ggplot2`. We will also use the `tibble` package for nicer data frames. We will use these packages:

```
library(tidygraph)
library(ggraph)
library(tibble)
```

To translate a tree, such as our search tree, into a graph we can plot with `ggraph`, we need to extract information about the edge structure. One approach is to map each node to a number and then create lists of `from` and `to` nodes. To do this, we can annotate the tree with node numbers. Because trees are immutable, we cannot simply add numbers to nodes—

we need to create a new tree that has these numbers. We can implement this with the following function:

```
node_number_annotate_tree <- function(tree, i = 1) {
  if (is_empty(tree)) {
    tree$dfn <- i
  } else {
    left <- node_number_annotate_tree(tree$left, i)
    right <- node_number_annotate_tree(tree$right, left$dfn + 1)
    tree$dfn <- right$dfn + 1
    tree$left <- left
    tree$right <- right
  }
  tree
}
```

Of course, making a separate copy of our data structure just to have a mapping from nodes to numbers is wasteful, but we do not plan to plot enormous trees, so it will not be a problem.

Once we have the node numbers, we can extract graph information. For the search tree, we want to know the values stored in nodes and the edge structure. The values we collect in a vector and the edge structure in two, a from and a to vector. It *is* possible to collect all the information we need in a purely functional way, but we will use the <<- operator and an imperative approach because the resulting code is easier to follow. In any case, we will only use this inside nested functions, and the function for extracting graph information will not have any side effects. We allocate a vector of the values in the nodes and the from and to vectors (there will be one less edge than there are nodes, except the root has an edge going to their parents). It is not easy to know for each node which index in the from and to it should use, because that depends on how many nodes we have seen earlier in the recursion through the tree. Therefore, we use a counter variable to keep track of that, and this one we also use the <<- operator to

handle between function invocations. We collect everything and return
it as a list with two nibble data frames: one for the nodes and one for the
edges:

```
extract_graph <- function(tree) {
  n <- tree$dfn
  values <- vector("numeric", length = n)
  from <- vector("integer", length = n - 1)
  to <- vector("integer", length = n - 1)
  edge_idx <- 1

  extract <- function(tree) {
    # we change the index so the root is number 1
    # that is easier
    i <- n - tree$dfn + 1
    values[i] <<- ifelse(is.na(tree$value), "", tree$value)

    if (!is_empty(tree)) {
      j <- n - tree$left$dfn + 1
      from[edge_idx] <<- i
      to[edge_idx] <<- j
      edge_idx <<- edge_idx + 1

      k <- n - tree$right$dfn + 1
      from[edge_idx] <<- i
      to[edge_idx] <<- k
      edge_idx <<- edge_idx + 1

      extract(tree$left)
      extract(tree$right)
    }
  }
```

```
  extract(tree)
  nodes <- tibble(1:n, value = values)
  edges <- tibble(from = from, to = to)
  list(nodes = nodes, edges = edges)
}
```

The nodes and edges we return from extract_graph can directly be provided to the tbl_graph from tidygraph to construct a structure that works with ggraph. We put the plotting code in the plot generic function for the unbalanced_search_tree class. It will not work *exactly* like a base graphics plot, because it is based on ggplot2 and therefore on the grid system, but we might as well call the function plot. The function will return a graphics object, though, so you need to print it to plot—just calling plot on a tree in the outermost scope will do that, though. The plotting function can look like this:

```
plot.unbalanced_search_tree <- function(x, ...) {
  x %>% node_number_annotate_tree %>%
    extract_graph %$% tbl_graph(nodes, edges) %>%
    mutate(leaf = node_is_leaf()) %>%
    ggraph(layout = "tree") +
    scale_x_reverse() +
    geom_edge_link() +
    geom_node_point(aes(filter = leaf),
                    size = 2, shape = 21, fill = "black") +
    geom_node_point(aes(filter = !leaf),
                    size = 10, shape = 21, fill = "white") +
    geom_node_text(aes(label = value), vjust = 0.4) +
    theme_graph()
}
```

In this plotting routine, we explicitly show the sentinel nodes. We could have left those out if we preferred. There are a lot of graphical parameters, such as sizes and shapes, that we might want to parameterize, but to keep it simple, I will leave that to you.

We can use it to illustrate trees constructed by inserting elements in different orders:

```r
tree <- empty_search_tree()
for (x in 1:7)
  tree <- insert(tree, x)
plot(tree) + ggtitle("Ordered")

tree <- empty_search_tree()
for (x in rev(1:7))
  tree <- insert(tree, x)
plot(tree) + ggtitle("Reverse")

tree <- empty_search_tree()
for (x in sample(1:7))
  tree <- insert(tree, x)
plot(tree) + ggtitle("Random")
```

The results are shown in Figure 3-12, which illustrates how the shape of the tree depends on the order in which we insert elements.

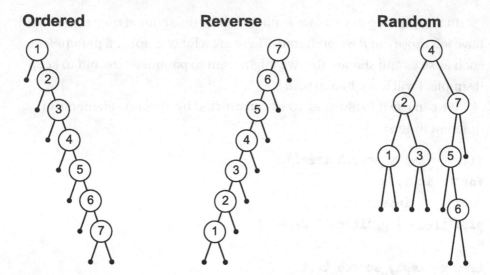

Figure 3-12. *Unbalanced search trees for different construction orders*

Lists and trees are the basic building blocks of all the data structures we will examine in this book, and the way we have written functions for updating them will be the way we approach all the data structures we see. The methods we saw in this chapter are not efficient in most cases—although the search tree is pretty good if the data we store in it is random—but the data structures that *are* efficient are built on the foundation we are familiar with now.

Random Access Lists

As a small example, before we move on to more traditional data structures, we will consider how to build random access lists, as described in Okasaki (1995a), using lists and trees. Random access lists behave like lists. We can extend them at the front, and we can access or remove the front element in constant time. However, they have two additional operations: we can access any element by its index, and we can change any element referenced by its index (but not delete elements at arbitrary positions).

The data structure from Okasaki (1995a) implements these two additional operations in $O\log(n)$ time.

The random access lists are constructed from fully balanced binary trees. With this data structure, we won't be implementing different versions, so I will not bother with classes here and only implement a bare-bones version of the data structures so that the actual structure can easily be seen and not be overshadowed by interface design. We implement trees as lists, and we will use NULL to indicate an empty tree. We can construct a tree like this:

```
ral_binary_tree <- function(value, left, right) {
  list(value = value, left = left, right = right)
}
```

Because we use trees to represent lists, we need a way of mapping positions in lists to nodes in the trees. In this data structure, we will map indices to the in-order position in the trees; this means that the first element in a tree is the value in the root, then follow all elements in the left subtree and then all elements in the right subtree. We choose this in-order mapping because we then have constant time access to the front of the list—the root of the tree.

To find the node in a tree that corresponds to a given index, we search recursively. Index 1 is always the root, so if we want to access this index, we have access to it right away. Otherwise, we have to search in either the left or the right subtree. If the tree has size n, then the root has 1 element, and each subtree has $(n - 1)/2$ elements, since all the trees we will manipulate are fully balanced. If we are looking for an index i larger than one, we know that we need to search in the left subtree if $i \leq (n-1)/2+1$ and that we need to search in the right subtree otherwise. In a recursive call, we then need to modify the index to skip past the root when searching to the left, or the root and the left subtree when searching to the right. Based on this algorithm, we can implement the lookup operation like this:

```
ral_is_leaf <- function(tree)
  !is.null(tree) && is.null(tree$left) && is.null(tree$right)

ral_tree_lookup <- function(tree, tree_size, idx) {
  if (idx == 1) return(tree$value)

  if (ral_is_leaf(tree)) # a leaf but idx is not one!
    if (idx > 1) stop("Index error in lookup")

  child_size <- (tree_size - 1) / 2
  if (idx <= child_size + 1)
    ral_tree_lookup(tree$left, child_size, idx - 1)
  else
    ral_tree_lookup(tree$right, child_size, idx - 1 -
      child_size)
}
```

Updating a tree is only slightly more involved. Locating the position where we need to change a value follows exactly the same search algorithm, of course, we just need to construct a new tree in recursive calls. Whenever we call recursively, we create a new tree from the existing values and trees, but modify either the left or the right subtree via a recursive call to the update operation:

```
ral_tree_update <- function(tree, tree_size, idx, value) {
  if (ral_is_leaf(tree)) {
    if (idx == 1)
      return(ral_binary_tree(value, NULL, NULL))
    # a leaf but idx is not one!
    stop("Index error")
  }
```

```
if (idx == 1) {
  ral_binary_tree(value, tree$left, tree$right)
} else {
  child_size <- (tree_size - 1) / 2
  if (idx <= child_size + 1) {
    ral_binary_tree(tree$value,
                    ral_tree_update(tree$left, child_size,
                    idx - 1, value),
                    tree$right)
  } else {
    ral_binary_tree(tree$value,
                    tree$left,
                    ral_tree_update(tree$right, child_size,
                                    idx - 1 - child_size,
                                    value))
  }
}
}
```

The trees are fully balanced, so the maximal depth of a tree of size n is $O(\log n)$, and both the lookup and the update operations thus run in this time.

Obviously, we cannot represent all lists with n elements in fully balanced trees. For any balanced tree with n nodes, n must be 2^k-1 for some k. Because of this, there is a close correspondence between fully balanced trees and skew binary numbers that we will exploit in this data structure. *Skew binary numbers* are positional numbers—meaning that the position of a digit determines its value, as you are used to with decimal and binary numbers—where a digit d at position k represent the number $d \times 2^{k+1}-1$. The digit d can be 0, 1, or 2, but is only allowed to be 2 if it is the least significant non-zero digit. The first six numbers are shown in decimal, binary and skew binary representations in the following table.

Decimal	Binary	Skew binary
0	0	0
1	1	1
2	10	2
3	11	10
4	100	11
5	101	12

There are two important properties of skew binary numbers that we will exploit for our random access lists: any number n can be represented in $O(\log n)$ digits (of which at most one is 2), and we can increment and decrement numbers in this representation in constant time. The first property should be obvious: we use fewer digits in skew binary numbers than in binary numbers, and for binary numbers, we only have $O(\log n)$ digits to represent the number n. Constant time increment and decrement can be achieved by representing only the non-zero digits as a linked list. We represent a number by a linked list where the elements in the list are the positions of the non-zero digits, where the digit 2 is represented by two elements with the same position, and we keep this list sorted in increasing position order, as illustrated in Figure 3-13.

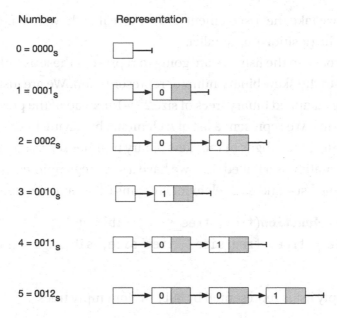

Figure 3-13. *Representation of skew binary numbers*

To increment a number, we first check whether the first two elements in the list contain the same position. If they do, we cannot prepend a position zero element because only the least significant non-zero digit can be 2. Instead, we can merge the two front elements into a single list element containing the next position in the number representation. Because $2\times(2^{k+1}-1)+1=2^{k+2}-1$, if the two front element contain the position k, then merging them this way increments the number n. Because only the two front elements can be duplicated, the new number we create this way doesn't invalidate the invariant—it might contain the same position as the next element in the list, but before we merge the front elements, that would be the only element containing that index. If the two front elements contain different positions, we can just add a zero-position element to the front of the list—this might change a 1 at position 0 to a 2, but that's fine because there was only one position 0 element to begin with.

To decrement, we do the reverse of the increment operation. If there is a zero-position element at the front of the list, we just remove that element.

CHAPTER 3   IMMUTABLE AND PERSISTENT DATA

Otherwise, we take the first element and replace it with two elements containing the position one smaller.

Getting back to the lists, we are going to represent these as linked lists similar to the skew binary number representation. We are just going to have fully balanced binary trees of size $2^{k+1}-1$ instead of the position k as list elements. We represent a list of n elements by having trees of sizes corresponding to the digits in the skew binary number representation of n. The representation is a linked list—we have a siblings pointer to the next element in the list—and each element contains a tree and the tree size:

```
ral_node <- function(tree, tree_size, siblings) {
  list(tree = tree, tree_size = tree_size, siblings = siblings)
}
```

The empty list is represented by NULL, as for empty trees:

```
ral_is_empty <- function(ral) is.null(ral)
```

We will keep the trees sorted such that the first elements in the list are in the first tree, the following in the second tree, and so on. Then, because the order of elements in individual trees is in-order, we can get the head of the list from the root of the first tree:

```
ral_head <- function(ral) {
  ral$tree$value
}
```

The other lists operations—appending an element to the front of the list, or removing the front element—follow the operations on skew binary numbers. For the cons operation, if the two front trees have different sizes, we just prepend a size 1 tree. If the two front trees have the same size, we construct a new tree with the new front element at the root, and the two original front trees as the left and right trees, respectively—the first tree must go to the left and the second to the right—and we replace the two front trees with the new tree. Taking the tail of a list is just the reverse of these operations: if the front tree has size 1, simply remove it, otherwise split the front tree into two and get rid of the root (Figure 3-14).

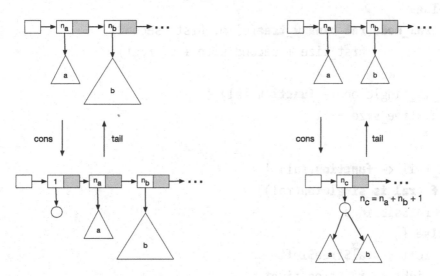

Figure 3-14. *List operations on random access lists*

The implementation just checks the cases and does the appropriate operations based on them:

```
ral_singleton_node <- function(value, siblings = NULL) {
  singleton_tree <- ral_binary_tree(value, NULL, NULL)
  ral_node(singleton_tree, 1, siblings)
}
ral_cons <- function(elem, ral) {
  if (ral_is_empty(ral) || ral_is_empty(ral$siblings))
    return(ral_singleton_node(elem, ral))

  first <- ral$tree
  first_size <- ral$tree_size
  second <- ral$siblings$tree
  second_size <- ral$siblings$tree_size
  rest <- ral$siblings$siblings
```

```r
  if (first_size < second_size)
    ral_singleton_node(elem, ral)
  else
    ral_node(ral_binary_tree(elem, first, second),
              first_size + second_size + 1, rest)
}
ral_is_singleton <- function(ral) {
  ral$tree_size == 1
}

ral_tail <- function(ral) {
  if (ral_is_singleton(ral))
    ral$siblings
  else {
    left <- ral$tree$left
    right <- ral$tree$right
    size <- (ral$tree_size - 1) / 2
    ral_node(left, size, ral_node(right, size, ral$siblings))
  }
}
```

What remains to be implemented are the lookup and update operations. We already have functions for looking up or updating a value in a tree, so we just need code for picking the right tree from the list of trees. This is a simple matter of keeping track of the index we are looking for. For the lookup, we iterate through the trees and check whether the index is less than the size of the current tree. If it is, we look into the tree. Otherwise, we decrease the index to reflect that we have moved past part of the list and then move to the next tree:

```r
ral_lookup <- function(ral, idx) {
  while (!is.null(ral)) {
    if (idx <= ral$tree_size)
```

```
    return(ral_tree_lookup(ral$tree, ral$tree_size, idx))
    idx <- idx - ral$tree_size
    ral <- ral$siblings
  }
  stop("Index out of bounds")
}
```

For the update operation, we are essentially doing the same thing as for the lookup operation, but because we need to construct a new list down to the tree we end up modifying, we cannot use a loop. We must do the search in a recursion that constructs a new list, so each recursive call must be wrapped in a construction of a new node:

```
ral_update <- function(ral, idx, value) {
  if (idx < 1) stop("Index out of bounds")
  if (idx <= ral$tree_size)
    ral_node(ral_tree_update(ral$tree, ral$tree_size, idx,
    value),
             ral$tree_size, ral$siblings)
  else
    ral_node(ral$tree, ral$tree_size,
             ral_update(ral$siblings, idx - ral$tree_size,
             value))

}
```

The way the data structure is updated is illustrated in Figure 3-15. Here, the index we must update is found in the third tree, so the new list will consist of three new elements and will then be followed by elements from the old list. The first two list elements just point to the first two original trees, whereas the third element points to a modified tree. In the figure, the original and updated trees are shown as disjoint, but in fact, they will share most of their structure; only elements on the path down to the node that is updated will differ between the two trees.

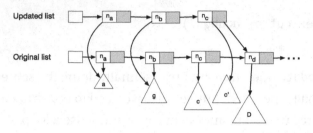

Figure 3-15. *Updating random access lists*

CHAPTER 4

Bags, Stacks, and Queues

In this chapter, we start building data structures that are more complex than simple linked lists and unbalanced trees, but still relatively simple compared to those in future chapters. We will consider three abstract data structures: the bag, the list, and the queue. The *bag* is just a set, but one where we don't value membership queries highly enough to make that operation efficient—we just want to be able to construct collections of elements that we can efficiently traverse later. The *stack* is a last-in-first-out collection of elements, where we can efficiently extract the last element we added to the stack. Finally, the *queue* is a first-in-first-out collection, where we can efficiently get at the first element we inserted into the queue.

The abstract data type operations we want to have for the three data structures are, for bags

```
is_empty <- function(x) UseMethod("is_empty")
insert <- function(x, elm) UseMethod("insert")
```

and sometimes

```
merge <- function(x, y) UseMethod("merge")
```

© Thomas Mailund 2017

T. Mailund, *Functional Data Structures in R*, https://doi.org/10.1007/978-1-4842-3144-9_4

where is_empty is the emptiness check we have used before, insert adds a single element to a bag, and merge combines two bags, like taking the union of two sets.

For stacks, we want the operations

```
is_empty <- function(x) UseMethod("is_empty")
push <- function(x, elm) UseMethod("push")
pop <- function(x) UseMethod("pop")
top <- function(x) UseMethod("top")
```

where push adds an element to the top of the stack, pop removes the top element, and top returns the top element.

Finally, for queues, we want the following operations:

```
is_empty <- function(x) UseMethod("is_empty")
enqueue <- function(x, elm) UseMethod("enqueue")
front <- function(x) UseMethod("front")
dequeue <- function(x) UseMethod("dequeu")
```

Here, enqueue inserts an element in the back of the queue, front returns the element at the front of the queue, and dequeue removes the element at the front of the queue.

Bags

Bags are probably the simplest data structures we can imagine. They are just collections of elements. You can add to them and merge two of them, but you cannot remove elements, and there is no natural order to them as there are with lists.

We can easily implement them using lists, where we already know how to insert elements in constant time and how to traverse the elements in a list. The merge operations, however, is a linear time function when lists are immutable.

Let us consider the list solution to bags. To work with a list version of
bags, we need to be able to create an empty bag and test for emptiness. We
can do this just as we did for creating lists:

```
bag_cons <- function(elem, lst)
  structure(list(item = elem, tail = lst),
            class = c("list_bag", "linked_list"))

bag_nil <- bag_cons(NA, NULL)
is_empty.list_bag <- function(x) identical(x, bag_nil)
empty_list_bag <- function() bag_nil
```

The only thing worth noticing here is that I made the class of the
elements in the bag-list both "list_bag" and "linked_lists". This
allows me to treat my bags as if they were lists because in every sense of
the word they actually are. We have to be a little careful with that, though,
because if we invoke methods that are set to return lists, then these will
not automatically be turned into bags just because we call them with a
bag. Using list functions is acceptable for queries, but problematic when
modifying bags. In any case, we only have two operations we need to
implement for updating bags, so we can handle that.

The simplest operation is insert where we can just put a new element
at the front of the list:

```
insert.list_bag <- function(x, elm) bag_cons(elm, x)
```

The merge operation involves more than the others. If we know that
the two bags we are merging contain disjoint sets of elements, we can
implement it by just concatenating the corresponding lists. We can reuse
the list_concatenate function from earlier, but we need to remember to
set the class of the result. This works as long as we never take the tail of
the result—that would give us a list and not a bag because of the way

list_concatenate works, but then, taking the tail of a bag is not really part of the interface to bags anyway. So we could implement merge like this:

```
merge.list_bag <- function(x, y) {
  result <- list_concatenate(x, y)
  class(result) <- c("list_bag", "linked_list")
  result
}
```

Because list concatenation is a linear time operation, bag merge is as well. We can, however, improve on this by using a tree to hold bags. We can exploit the fact that binary trees with n leaves have n–1 inner nodes, so if we put all our values in leaves of a binary tree, we can traverse it in linear time.

We can construct a binary tree bag with some boilerplate code for the empty tree like this:

```
bag_node <- function(elem, left, right)
  structure(list(item = elem, left = left, right = right),
            class = "tree_bag")

tree_bag_nil <- bag_node(NA, NULL, NULL)
is_empty.tree_bag <- function(x) identical(x, tree_bag_nil)
empty_tree_bag <- function() tree_bag_nil
```

Then, for inserting a new element, we create a leaf. If we try to insert the element into an empty tree we should just return the leaf—if we do not, we end up with non-binary nodes, and then the running time goes out the window. Otherwise, we just put the leaf to the left of a new root and the bag at the right. There is no need to keep the tree balanced because we don't plan to search in it or delete elements from it; for bags, we just want to be able to traverse all the elements:

```
insert.tree_bag <- function(x, elm) {
  element_leaf <- bag_node(elm, empty_tree_bag(),
  empty_tree_bag())
  if (is_empty(x)) element_leaf
  else bag_node(NA, element_leaf, x)
}
```

Figure 4-1 illustrates a tree bag and the insert operation. On the left is shown a bag containing the elements 1, 5, and 7, created with these operations:

```
x <- insert(insert(insert(empty_tree_bag(), 7, 5, 1)))
```

On the right, we see the situation after running

```
insert(x, 4)
```

The new element is added as the left tree of a new root, and the original x bag is the right subtree of the new root.

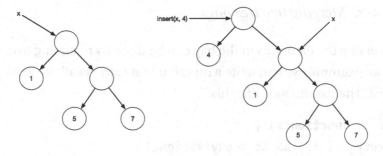

Figure 4-1. *Inserting elements in a tree bag*

Merging can now be done in constant time. We need to be careful not to create inner nodes with empty subtrees, but otherwise, we can just create a node that has the two bags we merge as its left and right subtrees:

```
merge.tree_bag <- function(x, y) {
  if (is_empty(x)) return(y)
  if (is_empty(y)) return(x)
  bag_node(NA, x, y)
}
```

Figure 4-2 illustrates a merge between two non-empty bags, x and y.

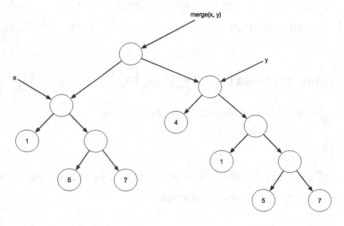

Figure 4-2. *Merging two tree bags.*

Traversing the elements in the bag can be done by recursing over the tree. As an example, we can write a function that extracts all the leaves as a linked list. That would look like this:

```
is_leaf <- function(x) {
  is_empty(x$left) && is_empty(x$right)
}
bag_to_list <- function(x, acc = empty_list()) {
  if (is_leaf(x)) list_cons(x$item, acc)
  else bag_to_list(x$right, bag_to_list(x$left, acc))
}
```

This implementation of merge also assumes that the bags we merge are disjoint sets. If they are not, we have more work to do. I am not aware of any general efficient solutions to this problem, if we don't want duplications, but two $n\log(n)$ algorithms are straightforward: we can sort the bags and merge them—this would take $n\log(n)$ for the sorting and then linear time for merging them—or we could simply put all the bag elements into a search tree and extract them from there. We can sort elements in this complexity using the heap data structure (discussed in Chapter 5), and we can construct and extract the unique elements in a search tree (discussed in Chapter 6).

Stacks

Stacks are collections of elements where we can insert elements and remove them again in a "last in, first out" order. Stacks are, if possible, even easier to implement using linked lists. Mostly, we just need to rename some of the functions; all the operations are readily available as list functions.

We need some boilerplate code again to get the type right for empty stacks:

```
stack_cons <- function(elem, lst)
  structure(list(item = elem, tail = lst),
            class = c("stack", "linked_list"))

stack_nil <- stack_cons(NA, NULL)
is_empty.stack <- function(x) identical(x, stack_nil)
empty_stack <- function() stack_nil
```

After that, we can just reuse list functions:

```
push.stack <- function(x, elm) stack_cons(elm, x)
pop.stack <- function(x) list_tail(x)
top.stack <- function(x) list_head(x)
```

That was pretty easy. Queues, on the other hand . . . those will require a bit more work.

Queues

Stacks are easy to implement because we can push elements onto the head of a list, and when we need to get them again, they are right there at the head. Queues, on the other hand, have a first-in-first-out semantics, so when we need to get the element at the beginning of a queue, if we have implemented the queue as a list, it will be at the *end* of the list, not the head. A straightforward implementation would, therefore, let us enqueue elements in constant time but get the front element and dequeue it in linear time.

There is a trick, however, for getting an amortized constant time operations queue. This means that the worst-case time usage for each individual operations will not be constant time, but whenever we have done *n* operations in total, we have spent time in $O(n)$. The trick is this: we keep track of two lists, one that represents the front of the queue and one that represents the back of the queue. The front of the queue is ordered such that we can get the front as the head of this list, and the back of the queue is ordered such that we can enqueue elements by putting them at the head of that list. From time to time we will have to move elements from the back list to the front lists; this we do whenever the front list is empty and we try to get the front of the queue or try to dequeue from the queue.

This means that some front or dequeue operations will take linear time instead of constant time, of course. Whenever we need to move elements from the back of the queue to the front, we need to copy and reverse the back of the queue list. On average, however, a linear number of operations take a linear amount of time. To see this, imagine that each enqueue operation actually takes twice as long as it really does. When we enqueue an element we spend constant time, so doubling the time is still constant time, but you can think of this doubling as paying for enqueueing an element *and* moving it from the back to the front. The first half of the time cost is paid right away when we enqueue the element; the second half you can think of as being put in a time bank. We are saving up time that we can later use to move elements from the back of the queue to the front. Because we put time in the bank whenever we add an element to the back of the queue, we always have enough in reserve to move the elements to the front of the queue later. Not all operations are constant time, but the cost of the operations are amortized over a sequence of operations, so on average they are.

Figure 4-3 illustrates the amortized time complexity. The solid line shows the number of list operations we have performed while doing a series of queue operations. We count one operation per enqueue function call and one operation per front and dequeue when we only modify the front list. When we move elements from the back list to the front list, we count the number of elements we move as well. The dashed line shows the time usage plus the number of operations we have in the bank. The line thus shows $2e+f+d$, where e is the number of enqueue operations, f is the number of front operations, and d is the number of dequeue operations. When we enqueue, we put value in the bank; that then pays for the expensive front or dequeue operations further down the line. A linear upper bound for the entire running time is simply two times the number of operations—shown as the dotted line.

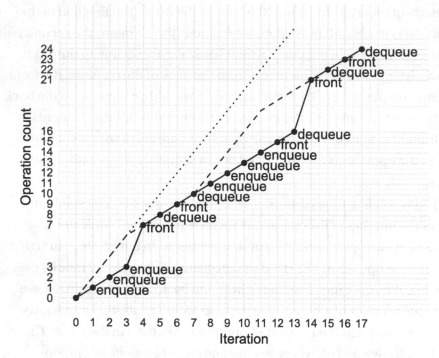

Figure 4-3. *Amortized linear bound on queue operations.*

So, we can implement a queue as two lists, but a problem presents itself now. Updating and querying the data structure are not completely separate operations any longer. If we try to get the front element of a queue, we might have to update the queue; if the front list is empty and the back list is not, we need to move all the elements from the back list to the front list before we can get the front element. This is something we will generally want to avoid; if querying a data structure also modifies it, we will need to return both the result of queries and the updated data structure on these operations, which breaks the clean interface and makes for some ugly code. It is, however, not unusual that we have data structures with good amortized time complexity, if not worst case complexity, that relies on modifying them when we query them. Implementing this queue solution gives me the opportunity to show you a general trick for handling queries that modify data structures in R: using environments that we *can* modify.

After that, I will show you a simpler solution for queues—we can actually extend the representation of queues a tiny bit to avoid the problem, but this isn't always possible, so it is worth knowing the general trick.

What we want, essentially, is for the `front` operation to have the side effect of moving elements from the back of the queue and to the front in case the front list is empty. With the `dequeue` operation, we don't have a problem; that operation returns an updated queue in any case. But `front` should only return the front of the list—any new queue it might need to construct isn't returned. If it were entirely impossible for R functions to have side effects, that would be the end of this strategy, but R is not a pure functional language, and some side effects are possible. We cannot modify data, but we can modify environments. We can change the binding of variables to data values. We can construct a queue that we can query and modify at the same time if we use an environment to hold values we need to update. We can do this explicitly using an environment object, or implicitly using closures. Of course, both environment-based solutions will not give us a persistent data structure. Because we do introduce side effects, we get a more traditional ephemeral data structure—but at the end of the chapter you will see how we get a persistent queue, so no worries.

Side Effects Through Environments

Because we can modify environments, we can create an environment object and use it as our queue. We can construct it like this, set the class so we can treat it as both an environment and a queue, and store the two lists in it:

```
queue_environment <- function(front, back) {
  e <- new.env(parent = emptyenv())
  e$front <- front
  e$back <- back
  class(e) <- c("env_queue", "environment")
  e
}
```

Here, we set the parent of the environment to the empty environment. If we didn't, the default would be to take the local closure as the parent. This wouldn't hurt us in this particular instance, but there is no need to chain the environment, so we don't.

Obviously, this isn't a purely functional data structure, and neither is it a persistent data structure, but it gets the job done.

We don't need a sentinel object to represent the empty queue with this representation. We can construct a queue with two empty lists, and we can check whether the two lists in the queue are empty:

```r
empty_env_queue <- function()
  queue_environment(empty_list(), empty_list())

is_empty.env_queue <- function(x)
  is_empty(x$front) && is_empty(x$back)
```

The operations on queues are relatively straightforward. When we add an element to the back of a queue, we just put it at the front of the back list:

```r
enqueue.env_queue <- function(x, elm) {
  x$back <- list_cons(elm, x$back)
  x
}
```

If the front list is empty, we need to replace it with the reversed back list and set the back list to empty—but otherwise, we just take the head of the front list:

```r
front.env_queue <- function(x) {
  if (is_empty(x$front)) {
    x$front <- list_reverse(x$back)
    x$back <- empty_list()
  }
  list_head(x$front)
}
```

Finally, to remove an element from the front of the queue, we just replace the front with the tail of the front list. If the front list is empty, though, we need to update it first, just as for the front function:

```
dequeue.env_queue <- function(x) {
  if (is_empty(x$front)) {
    x$front <- list_reverse(x$back)
    x$back <- empty_list()
  }
  x$front <- list_tail(x$front)
  x
}
```

Strictly speaking, we didn't have to return the updated queue in the enqueue and dequeue functions, since we are updating the environment that represents it directly. We should do it anyway, though, if we want to implement the interface we specified for the abstract data type. If we have those two functions return an updated queue, even when we are not implementing a persistent data structure, we can later replace the implementation with one that doesn't modify data but indeed implements queues as persistent structures.

Side Effects Through Closures

There is an alternative way of implementing a queue as an environment that does so with an implicit rather than an explicit environment. I don't personally find this approach as readable as using an explicit environment, but I have used the trick in a few situations where I do think it improves readability, and I have seen it used a few places, so I thought I might as well show it to you.

If we create *closures*—functions defined inside other functions—we get an implicit enclosing environment that we can use to update data. We can use such a closure to get a reference to a queue that we can update. We can

79

create a local environment that contains the front and the back lists and update these in closures, but just to show an alternative, I will instead keep a single queue object in the closure environments and update it by replacing it with updated queues when I need to. The queue object will look like this:

```r
queue <- function(front, back)
  list(front = front, back = back)
```

The closures will now work like this: I create some functions inside another function—a function that has a local variable that refers to a queue and returns the functions in a list. When functions need to modify the queue, they assign new versions of the queue to the variable in the enclosing environment using the `<<-` assignment operator. All the closures refer to the same variable, so they all see the updated version when they need to access the queue. I collect all the closures in a list that I return from the closure-creating function, with a class set to make it work with generic functions. The full implementation looks like this:

```r
queue_closure <- function() {
  q <- queue(empty_list(), empty_list())

  queue_is_empty <- function()
    is_empty(q$front) && is_empty(q$back)

  enqueue <- function(elm) {
    q <<- queue(q$front, list_cons(elm, q$back))
  }

  front <- function() {
    if (is_empty(q$front)) {
      q <<- queue(list_reverse(q$back), empty_list())
    }
    list_head(q$front)
  }
```

```
dequeue <- function() {
  if (is_empty(q$front)) {
    q <<- queue(list_tail(list_reverse(q$back)),
    empty_list())
  } else {
    q <<- queue(list_tail(q$front), q$back)
  }

  structure(list(is_empty = queue_is_empty,
                 enqueue = enqueue,
                 front = front,
                 dequeue = dequeue),
            class = "closure_queue")
}
```

The essential access and modification of the queue are the same as for the implementation with the explicit environment except that I need to use a different name for the function that tests whether the queue is empty. If I called that function is_empty, I would be shadowing the global generic function, and then I couldn't use it on the lists I use to implement the queue. In the list that I return, I can still call it is_empty, though.

When I need to create an empty queue, I just call the closure:

```
empty_queue <- function() queue_closure()
```

Now, to implement the generic functions for the queue interface, we just need to dispatch calls to the appropriate closures:

```
is_empty.closure_queue <- function(x) x$queue_is_empty()
enqueue.closure_queue <- function(x, elm) {
  x$enqueue(elm)
  x
}
```

```
front.closure_queue <- function(x) x$front()
dequeue.closure_queue <- function(x) {
  x$dequeue()
  x
}
```

We still need to return the modified queue—that is, the list of closures—to implement the abstract interface, but other than that, we just call the closures to modify the queue when needed.

As I mentioned, I find the implementation using an explicit environment simpler in this particular case, but the two implementations are equivalent, and you can use either depending on your taste.

A Purely Functional Queue

The only reason we couldn't make the queue data structure purely functional was that the accessor function front needed to get to the last element in the back list, which we cannot do in amortized constant time unless we modify the queue when the front list is empty. Well, it is only one element we need access to in special cases of calls to front, so we can make a purely functional—truly persistent—queue if we just explicitly remember that value in a way where we can get to it in constant time. We can make an extended version of the queue that contains the two lists, front and back, *and* the element we need to return if we call front on a queue when the front list is empty:

```
queue_extended <- function(x, front, back)
  structure(list(x = x, front = front, back = back),
            class = "extended_queue")
```

With this representation, we will require that we always satisfy the invariant that x refers to the last element in the back list when the list is not empty. If back is empty, we don't care what x is.

We don't need a sentinel object for this implementation of queues; testing whether a queue is empty can still be done just by testing if the two lists are empty. We can create an empty queue from two empty lists, and if they are both empty, we don't need to have any particular value for the last element of the back list—we already know it is empty, so we shouldn't try to get to the front of the queue in either case:

```
empty_extended_queue <- function()
  queue_extended(NA, empty_list(), empty_list())
```

```
is_empty.extended_queue <- function(x)
  is_empty(x$front) && is_empty(x$back)
```

When we add an element to the back of a queue, we now need to remember the value of the element if it is going to end up at the back of the back list. It will if the back queue is empty, so if back is empty, we remember the value we add in x; otherwise, we keep remembering the value we already stored in the queue:

```
enqueue.extended_queue <- function(x, elm)
  queue_extended(ifelse(is_empty(x$back), elm, x$x),
                 x$front, list_cons(elm, x$back))
```

When we need the front element of the queue, we are in one of two situations: if the front list is empty, we need to return the last element in back, which we have stored in x. If front is not empty, we can just return the head of that list:

```
front.extended_queue <- function(x) {
  if (is_empty(x$front)) x$x
  else list_head(x$front)
}
```

When we remove the front element of the queue, we should just remove the front element of front, except when front is empty. Then we should reverse back and put it at the front, and when we empty back, then x doesn't have any particular meaning any longer:

```
dequeue.extended_queue <- function(x) {
  if (is_empty(x$front))
    x <- queue_extended(NA, list_reverse(x$back), empty_list())
  queue_extended(x$x, list_tail(x$front), x$back)
}
```

Time Comparisons

Figure 4-4 compares the time usage of the three queue implementations in practice (I have not included the code for the experiments and plotting, as they are very similar to the code we have already seen). The two environment-based implementations run in roughly the same wall time, but there is an extra overhead in the functional implementation. This is caused by wrapping the newly constructed queues in a new structure each time—on top of wrapping lists in structures whenever we manipulate them.

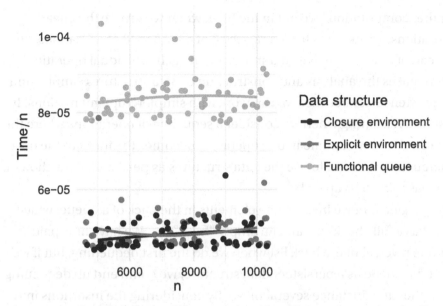

Figure 4-4. *Comparison of the running time for constructing and emptying the three implementations of queues*

We could speed up the implementation if we were willing to not wrap the implementation as an abstract data structure. We could do the same if we didn't represent the lists as objects and just used head and tail pointers. However, implementing a functional queue with the amortized complexity to avoid side effects isn't as useful as it might sound. The amortized complexity we achieve is excellent if we treat the queues as ephemeral data structures, but not actually as great for persistent data structures as you might think. The amortized analysis simply doesn't work if we want to use the queue as a persistent data structure.

Amortized Time Complexity and Persistent Data Structures

When we work with data structures with amortized complexity, we imagine that some of the cheap operations actually cost a little more than they do, and then we can afford some of the more expensive operations for the "computation" we put in the bank when we invoke the cheap operations. Thinking of the average cost of a sequence of operations, instead of having to make guarantees about each individual operation, often makes the analysis and construction of data structures simpler and quite often faster because we can work with simpler implementations. It is sound reasoning when we consider a sequence of operations where the data structure we update in one operation becomes the input to the next operation. If we want to use the data structures as persistent data, though, the reasoning falls apart.[1]

Imagine that we insert some elements in the back of a queue, which now has a full "back" list and an empty "front" list. We have now paid for one reversal of the back list when we do the first dequeuing, but if we treat the queue as a persistent data structure, we could end up dequeuing from the same instance several times. By considering the insertions into the queue as twice as expensive as they actually are—as we did in the analysis—we can pay for the first dequeuing, reversing the list and all, but it only pays for the *first* reversal of the back list. If the queue is persistent, there is nothing that prevents us from calling a dequeue operation on the same queue several times. Each call to dequeue will be an expensive operation, linear in the length of the back list, but the savings we put in the bank when we inserted elements in the queue only pays for the first one. The amortized analysis is really only valid if we treat the structure as ephemeral.

---

[1]Just to avoid any confusion here: a data structure can be immutable, so manipulation of it has no side effects, and still not give us a persistent data structure when we analyze its amortized complexity. The environment-based queues we have implemented here are both mutable and ephemeral data structures because accessing them incurs side effects. The functional queue is immutable—it has no side effects—but the amortized analysis forces us to treat it as ephemeral. If we only cared about its worst-case performance, we could treat it as persistent.

The queue implementations we have seen here are reasonably fast if we treat them as ephemeral, but if we want persistent queues, the amortized analysis doesn't apply. It *is* possible to combine amortized analysis and persistent data structures, but it requires a slightly different way of thinking. The problem with amortized analysis is that we have to save up ahead of time to pay for later expensive operations. And with persistent data structures, we risk spending the savings more than once. Instead, it is possible, through lazy evaluation, to leave the data structure in a state where it has unevaluated operations that must be paid for in some future operations—instead of savings, we have a debt. Any future series of operations we do on a data structure in such a state will have to pay off the depth before we can see the effect of the operation we delayed. We return to an example of this in the final section of this chapter.

Double-Ended Queues

A *double-ended queue*, also known as a *deque*, is a queue where you can add and remove elements from both ends of it. The operations on a double-ended queue, as an abstract data structure, would be these:

```
enqueue_front <- function(x, elm) UseMethod("enqueue_front")
enqueue_back <- function(x, elm) UseMethod("enqueue_back")

front <- function(x) UseMethod("front")
back <- function(x) UseMethod("back")

dequeue_front <- function(x) UseMethod("dequeue_front")
dequeue_back <- function(x) UseMethod("dequeue_back")
```

We can implement a double-ended queue using two lists, a front and a back list, just as we did with queues. We can readily add elements to both the front and the back by putting new elements at the head of the two lists. If we want to get the front or back of the elements, or if we want to remove the front or the back element, we can just access the head of the two lists, as long as they are not empty. If they are empty, though, we have to be a little more careful.

The problem with just following the same procedure as we did for queues is that if we reverse the front or back list whenever the other list is empty, we might end up with linear time operations when we switch between removing elements from the front and from the back. This is something we want to avoid, and we can do that with a small modification to the procedure: instead of reversing and moving an entire list, when the other is empty, we only move half the other list. We will end up with the same amortized complexity as for the queues—which means that as long as we implement double-ended queues as ephemeral data structures, we get constant time operations.

The complexity analysis is a little more involved. We can still think of the insertion operations as being slightly more expensive, taking one "computation coin" to execute and then putting one "computation coin" in the bank, but the bank holdings we now have to think of as the difference between the length of the front list and the back list. If you are familiar with amortized analysis, you should think of a potential function that is the absolute difference between the length of the front list and the back list. To keep it simple here, we will still think in terms of putting computation coins in a bank that we can then use to pay for expensive operations later on. It is just that the coin we put in the bank when we insert an element won't necessarily pay for all the times that an individual element gets moved between the lists. The same element might be moved between the lists more than once, which puts that element in debt, but we can recover that debt by using coins that are paid for by other elements.

If we let f denote the number of elements in the front list, and b denote the number of elements in the back list, then we want the invariant to be that we have abs($f-b$) in the bank to pay for future operations. If we insert an element in the shorter of the two lists, we might think of inserting an extra computation coin in the bank, but we don't really need to to satisfy the invariant, so it is just extra money we won't need. If we insert an element in the longer list, though, we need to pay an extra coin for the increased difference between the list lengths. That is okay—it is still a constant time operation even if it is twice as expensive as just inserting an element.

If one of the lists is empty, however, and we need to get an element from it, we need to move some elements, and we need to take the cost of that out of the bank. If the invariant is satisfied, though, we have as many coins in the bank as the length of the non-empty list. If we move half the elements in that list to the other list, we spend half the savings we have in the bank, but what remains is still enough to pay for the abs(**f–b**) invariant to be true. If the two lists end up being exactly the same length, we only require that there be a non-negative amount of coins in the bank—which will be true if we only spend half the coins there—and if the lists are one off, we need one remaining coin, which again will be true if we only spend half the coins there. So we are good.

To implement double-ended queues with a constant time (amortized) operation complexity, we just need to move half the lists when we need to move anything, rather than the entire lists.

To implement double-ended queues, we need two operations: get the first half of a list, and get the second half of a list.

If we know how many elements are in half the list, we can use these two functions for this:

```
list_get_n_reversed <- function(lst, n) {
  l <- empty_list()
  while (n > 0) {
    l <- list_cons(list_head(lst), l)
    lst <- list_tail(lst)
    n <- n - 1
  }
  l
}
```

```
list_drop_n <- function(lst, n) {
  l <- lst
  while (n > 0) {
    l <- list_tail(l)
    n <- n - 1
  }
  l
}
```

Both functions will do their work in time n, so we should be able to move elements from one list to the other with the time we have, if we let n be half the length of the list we move elements from.

We can get the list length like this:

```
list_length <- function(lst) {
  n <- 0
  while (!is_empty(lst)) {
    lst <- lst$tail
    n <- n + 1
  }
  n
}
```

We can then implement the double-ended queue like this:

```
deque_environment <- function(front, back) {
  e <- new.env(parent = emptyenv())
  e$front <- front
  e$back <- back
  class(e) <- c("env_deque", "environment")
  e
}
```

```r
empty_env_deque <- function()
  deque_environment(empty_list(), empty_list())

is_empty.env_deque <- function(x)
  is_empty(x$front) && is_empty(x$back)

enqueue_back.env_deque <- function(x, elm) {
  x$back <- list_cons(elm, x$back)
  x
}
enqueue_front.env_deque <- function(x, elm) {
  x$front <- list_cons(elm, x$front)
  x
}

front.env_deque <- function(x) {
  if (is_empty(x$front)) {
    n <- list_length(x$back)
    x$front <- list_get_n_reversed(x$back, ceiling(n))
    x$back <- list_drop_n(x$back, ceiling(n))
  }
  list_head(x$front)
}
back.env_deque <- function(x) {
  if (is_empty(x$back)) {
    n <- list_length(x$front)
    x$back <- list_get_n_reversed(x$front, ceiling(n))
    x$front <- list_drop_n(x$front, ceiling(n))
  }
  list_head(x$back)
}
```

```
dequeue_front.env_deque <- function(x) {
  if (is_empty(x$front)) {
    n <- list_length(x$back)
    x$front <- list_get_n_reversed(x$back, ceiling(n))
    x$back <- list_drop_n(x$back, ceiling(n))
  }
  x$front <- list_tail(x$front)
  x
}
dequeue_back.env_deque <- function(x) {
  if (is_empty(x$back)) {
    n <- list_length(x$front)
    x$back <- list_get_n_reversed(x$front, ceiling(n))
    x$front <- list_drop_n(x$front, ceiling(n))
  }
  x$back <- list_tail(x$back)
  x
}
```

If you are a little bit uncomfortable now, you should be. If you are not, I want you to look over the solution we have so far and think about where I might have been cheating you a little.

If you don't see it yet, I will give you a hint: we have half the length of the list we are taking elements from to spend from the bank, but are we using more than that?

Do you see it now? We are okay in nearly all the operations we are doing, but how do we get the length of the list?

It is unpleasantly easy to get it *almost* right when you implement data structures, but the slightest errors can hurt the performance, so I left a little trap in this data structure to keep you on your toes. What we have right now is *almost* right, but we do spend too much time in reversing half of the lists. We only have coins in the bank for taking half a list, but to get the length of the list, we spend the full list length in coins. To figure out how long a list is, the implementation I showed you before runs through the entire list. That is too much for our analysis to be correct. If you take the current implementation and experiment with it to measure the running time, you will see this—which is why I encourage you always to test your implementations with measurements and not just rely on analyses. It doesn't matter how careful we are in the design and analysis of a data structure, if we get a slight implementation detail wrong, it all goes out the window. Always check whether the performance you expect from your analysis is actually correct when measured against wall time.

The solution to this problem is simple enough. If we can figure out the lengths of the two lists in constant time, we can also move half of them in the allotted time. The simplest way to know the length of the lists when we need it is simply to keep track of it. So we can add the lengths of the lists to the double-ended queue as extra information:

```r
deque_environment <- function(front, back,
                              front_length, back_length) {
  e <- new.env(parent = emptyenv())
  e$front <- front
  e$back <- back
  e$front_length <- front_length
  e$back_length <- back_length
  class(e) <- c("env_deque", "environment")
  e
}

empty_env_deque <- function()
  deque_environment(empty_list(), empty_list(), 0, 0)
```

The test for emptiness doesn't have to change—there we just check whether the lists are empty, and if we are correct in keeping track of the lengths, we will be correct there as well.

Whenever we add an element to a list, we have to add one to the length bookkeeping as well:

```
enqueue_back.env_deque <- function(x, elm) {
  x$back <- list_cons(elm, x$back)
  x$back_length <- x$back_length + 1
  x
}
enqueue_front.env_deque <- function(x, elm) {
  x$front <- list_cons(elm, x$front)
  x$front_length <- x$front_length + 1
  x
}
```

Now, when modifying the lists, we need to update the lengths as well. That is used in several places, so we can implement two helper functions to keep track of it, like this:

```
move_front_to_back <- function(x) {
  n <- list_length(x$front)
  m <- ceiling(n)
  x$back <- list_get_n_reversed(x$front, m)
  x$front <- list_drop_n(x$front, m)
  x$back_length <- m
  x$front_length <- n - m
}

move_back_to_front <- function(x) {
  n <- list_length(x$back)
  m <- ceiling(n)
  x$front <- list_get_n_reversed(x$back, m)
  x$back <- list_drop_n(x$back, m)
  x$front_length <- m
  x$back_length <- n - m
}
```

Then, we can update the operations like this:

```
front.env_deque <- function(x) {
  if (is_empty(x$front)) move_back_to_front(x)
  list_head(x$front)
}
back.env_deque <- function(x) {
  if (is_empty(x$back)) move_front_to_back(x)
  list_head(x$back)
}

dequeue_front.env_deque <- function(x) {
  if (is_empty(x$front)) move_back_to_front(x)
  x$front <- list_tail(x$front)
  x
}
dequeue_back.env_deque <- function(x) {
  if (is_empty(x$back)) move_front_to_back(x)
  x$back <- list_tail(x$back)
  x
}
```

Now we only spend time moving elements linearly in the number of elements we move, and then the amortized analysis invariant is satisfied.

Lazy Queues

It is possible to implement worst-case $O(1)$ operation queues as well as amortized queues by exploiting lazy evaluation. In the following, I will describe an implementation of these following Okasaki (1995b).

Because lazy evaluation is not the natural evaluation strategy in R, except for function parameters that *are* lazily evaluated, we have to implement it ourselves. This adds some overhead to the data structure in the form of extra function evaluations, so this implementation cannot compete with the previous data structures we have seen when it comes to speed, but this queue can be used persistently.

Implementing Lazy Evaluation

Expressions in R are evaluated immediately except for expressions that are passed as parameters to functions. This means that an expression such as

```
1:10000
```

immediately creates a vector of ten thousand elements. However, if we write a function like this:

```
f <- function(x, y) x
```

where we don't access y, and call it with parameters like these

```
f(5, 1:10000)
```

the vector expression is never evaluated.

We can thus wrap expressions we don't want to evaluate just yet in *thunks*, functions that do not take any arguments but evaluate an expression. For example, we could write a function like this:

```
lazy_thunk <- function(expr) function() expr
```

It takes the expression expr and returns a thunk that will evaluate it when we call it. It will only evaluate it the first time, though, because R remembers the values of such parameters once they are evaluated. We can see this in the following code:

```
library(microbenchmark)
microbenchmark(lazy_vector <- lazy_thunk(1:100000), times = 1)
## Unit: microseconds
```

```
##                                   expr    min    lq
##   lazy_vector <- lazy_thunk(1:1e+05) 8.244 8.244
##    mean median    uq   max neval
##   8.244  8.244 8.244 8.244       1
microbenchmark(lazy_vector()[1], times = 1)
## Unit: microseconds
##                  expr     min      lq    mean  median
##   lazy_vector()[1] 272.457 272.457 272.457 272.457
##       uq     max neval
##   272.457 272.457     1
microbenchmark(lazy_vector()[1], times = 1)
## Unit: microseconds
##                expr   min    lq  mean median    uq
##   lazy_vector()[1] 4.784 4.784 4.784  4.784 4.784
##    max neval
##   4.784     1
```

The construction of the vector is cheap because the vector expression
isn't actually evaluated when we construct it. The first time we access
the vector, though—and notice that we need to evaluate lazy_vector as
a thunk to get to the expression it wraps—we will have to construct the
actual vector. This is a relatively expensive operation, but the second time
we access it, it is already constructed, and we will have a cheap operation.

We have to be a little careful with functions such as lazy_thunk. The
expr parameter will be evaluated when we call the thunk the first time,
and it will be evaluated in the calling scope where we constructed the
thunk but as it looks at the time when we evaluate the thunk. If it depends
on variables that have been modified since we created the thunk, the
expression we get might not be the one we want. The function force
is typically used to alleviate this problem. It forces the evaluation of a

parameter, so it evaluates the values that match the expression at the time the thunk is constructed. This won't work here, though—we implement the thunk exactly because we do not want the expression evaluated.

If we are careful with how we use thunks and make sure that we give them expressions where any variables have already been forced, though, we can exploit lazy evaluation of function parameters to implement lazy expressions.

Lazy Lists

Now, let us implement lazy lists to see how we can use lazy evaluation. We are only going to use the lazy lists wrapped in a queue, so I won't construct classes with generic functions, to avoid the overhead we get there and to simplify the implementation. To avoid confusion with the function names we use with the abstract data types, I will use different names for constructing and accessing lists: I will use terminology from Lisp-based languages and use cons for constructing lists, car for getting the head of a list, and cdr for getting the tail of a list. The names do not give much of a hint at what the functions do—they are based on IBM 704 assembly code—but they are common in languages based on Lisp, so you are likely to run into them if you study functional programming, and you might as well get used to them.

We will make the following invariant for lazy lists: a list is always a thunk that either returns NULL when the list is empty, or a structure with a car and a cdr field. We implement construction and access to lists like this:

```
nil <- function() NULL
cons <- function(car, cdr) {
  force(car)
  force(cdr)
  function() list(car = car, cdr = cdr)
}
```

```
is_nil <- function(lst) is.null(lst())
car <- function(lst) lst()$car
cdr <- function(lst) lst()$cdr
```

This is very similar to how we have implemented the linked lists we have used so far, except that we do not use polymorphic functions and we use NULL for the empty list instead of a sentinel.

We can take any of the functions we have written for linked lists and make them into lazy lists by wrapping what they return in a thunk. Take, for example, list concatenation. Using the functions for lazy lists, the implementation we have would look like this:

```
concat <- function(l1, l2) {
  rev_l1 <- nil
  while (!is_nil(l1)) {
    rev_l1 <- cons(car(l1), rev_l1)
    l1 <- cdr(l1)
  }
  result <- l2
  while (!is_nil(rev_l1)) {
    result <- cons(car(rev_l1), result)
    rev_l1 <- cdr(rev_l1)
  }
  result
}
```

This function simply does what the previous concatenation function did, but with lazy lists. It is not lazy itself, though. It takes linear time to execute and does the reversal right away.

We can experiment with it by constructing some long lists with this helper function:

```
vector_to_list <- function(v) {
  lst <- nil
  for (x in rev(v)) lst <- cons(x, lst)
  lst
}

l1 <- vector_to_list(1:100000)
l2 <- vector_to_list(1:100000)
```

We see that the concatenation operation is slow:

```
microbenchmark(lst <- concat(l1, l2), times = 1)
## Unit: seconds
##                        expr      min       lq      mean
##   lst <- concat(l1, l2) 1.669522 1.669522 1.669522
##    median       uq      max neval
##   1.669522 1.669522 1.669522      1
```

The operation takes more than a second to complete. Accessing the list after we have concatenate l1 and l2, however, is fast:

```
microbenchmark(car(lst), times = 1)
## Unit: microseconds
##        expr  min   lq mean median   uq  max neval
##   car(lst) 20.1 20.1 20.1   20.1 20.1 20.1      1
microbenchmark(car(lst), times = 1)
## Unit: microseconds
##        expr   min    lq  mean median    uq   max
##   car(lst) 9.134 9.134 9.134  9.134 9.134 9.134
##   neval
##        1
```

These operations run in microseconds, and because we are not delaying any operations, we spend the same time both times we call `car` on lst.

We can now try slightly modifying `concat` to delay its evaluation by wrapping its return value in a thunk. We need to `force` its parameters to avoid the usual problems with them referring to variables in the calling environment, but we can wrap the concatenation in a thunk after that:[2]

```
concat <- function(l1, l2) {
  do_cat <- function(l1, l2) {
    rev_l1 <- nil
    while (!is_nil(l1)) {
      rev_l1 <- cons(car(l1), rev_l1)
      l1 <- cdr(l1)
    }
    result <- l2
    while (!is_nil(rev_l1)) {
      result <- cons(car(rev_l1), result)
      rev_l1 <- cdr(rev_l1)
    }
    result
  }
  force(l1)
  force(l2)
  lazy_thunk <- function(lst) function() lst()
  lazy_thunk(do_cat(l1, l2))
}
```

---

[2]I will use an explicit function, `lazy_thunk`, to wrap operations in this chapter. You could equally well just return an anonymous function, but you would have to remember to evaluate the body as a function to make it behave like a list. So, instead of returning lazy_thunk(do_cat(l1,l2)) in the `concat` function, we could return function() do_cat(l1,l2)().

Now, the concatenation is a fast operation:

```
microbenchmark(lst <- concat(l1, l2), times = 1)
## Unit: microseconds
##                      expr    min      lq    mean
##   lst <- concat(l1, l2) 11.153 11.153 11.153
##   median       uq     max neval
##   11.153 11.153 11.153      1
```

The first time we access the list, though, we pay for the concatenation, but only the first time:

```
microbenchmark(car(lst), times = 1)
## Unit: seconds
##       expr      min       lq     mean   median
##   car(lst) 1.304761 1.304761 1.304761 1.304761
##         uq      max neval
##   1.304761 1.304761      1
microbenchmark(car(lst), times = 1)
## Unit: microseconds
##       expr   min    lq  mean median    uq    max
##   car(lst) 15.29 15.29 15.29  15.29 15.29 15.29
##   neval
##       1
```

We still haven't achieved much by doing this. We have just moved the cost of concatenation from the concat call to the first time we access the list. If we abandon the loop version of the concatenation function, though, and go back to the recursive version, we can get a simpler version where all operations are constant time:

```
concat <- function(l1, l2) {
  force(l1)
  force(l2)
```

```
if (is_nil(l1)) l2
else {
  lazy_thunk <- function(lst) function() lst()
  lazy_thunk(cons(car(l1), concat(cdr(l1), l2)))
}
}
```

In this function, we don't actually need to force l1 because we directly use it, but just for consistency I will always force the arguments in functions that return thunks.

We abandoned this version of the concatenation function because we would recurse too deeply on long lists, but by wrapping the recursive call in a thunk that we evaluate lazily, we do not have this problem. When we evaluate the result we get from calling this concat, we only do one step of the concatenation recursion. Now concatenation is a relatively fast operation—it needs to set up the thunk, but it doesn't do any work on the lists:

```
microbenchmark(lst <- concat(l1, l2), times = 1)
## Unit: microseconds
##                     expr    min     lq   mean
##   lst <- concat(l1, l2) 19.118 19.118 19.118
## median    uq    max neval
##   19.118 19.118 19.118     1
```

The first time we access the concatenated lists we need to evaluate the thunk. This is fast compared to actually concatenating them and is a constant time operation:

```
microbenchmark(car(lst), times = 1)
## Unit: microseconds
##      expr    min     lq   mean median     uq
##   car(lst) 50.398 50.398 50.398 50.398 50.398
##      max neval
##   50.398     1
```

Subsequent access to the head of the list is even faster. Now the thunk has already been evaluated, so we just access the list structure at the head:

```
microbenchmark(car(lst), times = 1)
## Unit: microseconds
##       expr    min      lq   mean median     uq
##   car(lst) 19.423 19.423 19.423 19.423 19.423
##       max neval
##    19.423     1
microbenchmark(car(lst), times = 1)
## Unit: microseconds
##       expr   min     lq  mean median    uq    max
##   car(lst) 8.928 8.928 8.928  8.928 8.928 8.928
##   neval
##       1
```

If we could do the same thing with list reversal, we would have a queue with constant time worst-case operations right away, but unfortunately we cannot. We could try implementing it like this:

```
reverse <- function(lst) {
  r <- function(l, t) {
    force(l)
    force(t)
    if (is_nil(l)) t
    else {
      lazy_thunk <- function(lst) function() lst()
      lazy_thunk(r(cdr(l), cons(car(l), t)))
    }
  }
  r(lst, nil)
}
```

That, however, just constructs a lot of thunks that when we evaluate the first—which we do at the end of the function—calls the entire recursion. Now both reversing the list and accessing it will be slow operations:

```
l <- vector_to_list(1:500)
microbenchmark(lst <- reverse(l), times = 1)
## Unit: microseconds
##                  expr    min      lq   mean median
##   lst <- reverse(l) 17.776 17.776 17.776 17.776
##        uq    max neval
##   17.776 17.776      1
microbenchmark(car(lst), times = 1)
## Unit: milliseconds
##      expr      min       lq     mean   median
##   car(lst) 3.365896 3.365896 3.365896 3.365896
##        uq      max neval
##   3.365896 3.365896      1
```

It is even worse than that. Because we are reversing the list recursively, we will recurse too deeply for R when we call the function on a long list.

Wrapping the recursion in thunks may make it seem as if we are not recursing, but we are constructing thunks that, when evaluated, will call the computation all the way to the end of the list. We are really just implementing an elaborate version of the recursive reversal we had earlier.

We could make accessing the list cheap by wrapping the reversal in a thunk, but because we cannot get the head of a reversed list without going to the end of the original list, we cannot make reversal into a constant time operation. The best we can do is to use the iterative reversal from earlier and wrap it in a thunk, so we at least get the reversal call as a cheap operation:

```
reverse <- function(lst) {
  do_reverse <- function(lst) {
    result <- nil
```

```
  while (!is_nil(lst)) {
    result <- cons(car(lst), result)
    lst <- cdr(lst)
  }
  result
}
force(lst)
lazy_thunk <- function(lst) {
  function() lst()
}
lazy_thunk(do_reverse(lst))
}
```

With this implementation, setting up the reversal is cheap—the first time we access the list we pay for it, but subsequent access is cheap again:

```
l <- vector_to_list(1:10000)
microbenchmark(lst <- reverse(l), times = 1)
## Unit: microseconds
##                   expr     min      lq    mean median
##    lst <- reverse(l) 11.879  11.879  11.879 11.879
##        uq     max neval
##    11.879 11.879      1
microbenchmark(car(lst), times = 1)
## Unit: milliseconds
##       expr     min       lq     mean   median
##    car(lst) 38.24003 38.24003 38.24003 38.24003
##          uq      max neval
##    38.24003 38.24003      1
microbenchmark(car(lst), times = 1)
```

```
## Unit: microseconds
##      expr   min    lq  mean median    uq   max
##  car(lst) 9.584 9.584 9.584  9.584 9.584 9.584
## neval
##      1
```

Because reversing the back list of a queue is the expensive operation, and we cannot get away from that with lazy lists, it seems like we are not getting far, but we did gain a little. It is now possible to cheaply set up a reversal that we just have to pay for later. This we can use to construct a lazy queue where we consider this reversal a debt that must be paid off by all future users that plan to see the effect of it. Because we memorize lazily evaluated expressions, only one of them will, but because all will have to pay the debt, we guarantee that any amortized analysis gives us time bounds that also work if we treat the queues as persistent.

Amortized Constant Time, Logarithmic Worst-Case, Lazy Queues

Following Okasaki (1995b) we will work our way up to a queue solution with worst-case constant time operations by first implementing a version that has constant time amortized operations, even when used as a persistent data structure, and where the worst-case time usage for any operation is logarithmic in the queue length. This is an improvement from the previous queues on both accounts, at least if we use the implementation as a persistent queue or if we need it to be fast for each operation—for example, if we use it in interactive code.

We represent the queue as a front and back list as before, and we have to keep track of the list lengths in this data structure as well. In this case

107

we will use lazy lists, but we will get to that. The constructor for the queue looks like this:

```
lazy_queue <- function(front, back, front_length, back_length)
{
  structure(list(front = front, back = back,
                 front_length = front_length,
                 back_length = back_length),
            class = "lazy_queue")
}
```

We can construct an empty queue, and check for emptiness, like this:

```
empty_lazy_queue <- function() lazy_queue(nil, nil, 0, 0)
is_empty.lazy_queue <- function(x)
  is_nil(x$front) && is_nil(x$back)
```

We will have the following invariant for the queue: the back list can at most be one longer than the front list. Whenever the back list grows larger than the front list, we are going to move the elements in it to the front queue, but we will do so lazily.

The idea behind the approach is this: if we lazily reverse and concatenate when the front list is longer than the back list, then any future use of the queue that actually invokes the reversal will have had to first dequeue enough elements from the front list to pay for the reversal. Unlike the previous queue implementations, where the amortized analysis relies on earlier enqueue operations paying for an expensive reversal, this implementation relies on future users invoking enough cheap dequeue operations before they can invoke an expensive one. We do not rely on the past having saved up for the expensive operation; we have put a debt on the queue—the expensive reversal—but any future user who wants to use the reversal will first have had to pay off the debt. Only one future user will actually invoke the expensive operation, all other future users might have paid off on the debt, but that is fine. You cannot spend your saving more than once, but there is no problem with paying off a debt more than once. This now works as a persistent data structure.

The implementation of the queue is based on a "rotate" function that combines concatenation and reversal. The function looks like this:

```
rot <- function(front, back, a) {
  force(front)
  force(back)
  force(a)
  if (is_nil(front)) cons(car(back), a)
  else {
    lazy_thunk <- function(lst) function() lst()
    lazy_thunk(cons(car(front),
                    rot(cdr(front), cdr(back), cons(car(back),
                        a)))))
  }
}
```

It operates on three lists: the front list, the back list, and an accumulator. The idea behind the rotation function is that it concatenates the front list to the back list in a lazy recursion, just as the earlier concatenation function did, but at the same time it reverses the back list one step at a time. We will call it whenever the front list is one element shorter than the back list. For each step in the concatenation, we also handle one step of the reversal. If the front list is empty, we put the first element of the back list in front of the accumulator. The queue invariant guarantees that if the front list is empty, the back queue only contains a single element. The recursive call, wrapped in a thunk, puts the first element of the front list at the beginning of a list and then makes the continuation of the list another rotation call.

To make sure that we call the rotate function whenever we need to, to satisfy the invariant, we wrap all queue construction calls in the following function:

```
make_q <- function(front, back, front_length, back_length) {
  if (back_length <= front_length)
    lazy_queue(front, back, front_length, back_length)
  else
    lazy_queue(rot(front, back, nil), nil,
               front_length + back_length, 0)
}
```

Its only purpose is to call rotate when we need to. Otherwise, it just constructs a queue.

The implementation of the queue abstract interface is relatively straightforward once we have these two functions:

```
enqueue.lazy_queue <- function(x, elm)
  make_q(x$front, cons(elm, x$back),
         x$front_length, x$back_length + 1)

front.lazy_queue <- function(x) car(x$front)

dequeue.lazy_queue <- function(x)
  make_q(cdr(x$front), x$back,
         x$front_length - 1, x$back_length)
```

When we enqueue an element, we add it to the back list; when we need the front element, we get it from the front list—which cannot be empty unless the entire queue is empty; and when we dequeue an element, we just shorten the front list. All this is wrapped in calls to make_q that calls rot when needed.

We can see how the queue works in action by going through a few operations and see what the lists look like after each. We start with making an empty queue. It will have two empty lists:

```
q <- empty_lazy_queue()
```

```
Front: nil
Back: nil
```

If we now add an element, we will prepend it to the back list and then call make_q that will call rot because the back list is longer than the front list. In rot we see that the front list is empty, so we have the base case where we just return the singleton list containing the element in the back list:

```
q < enqueue(q, 1)
```

```
Front: cons(1, nil)
Back: nil
```

If we add a second element, we will see in make_q that the back and front lists have the same length, so we just prepend the element to the back list:

```
q < enqueue(q, 2)
```

```
Front: cons(1, nil)
Back: cons(2, nil)
```

Inserting the third element, we get the first real rotation. We prepend 3 to the back list, so it now contains the sequence (3,2) while the front list contains (1), and we then construct a list with the head 1 and the tail the rotation of nil, 2 and 3, but wrap that in a thunk:

```
q < enqueue(q, 3)
```

```
Front: lazy_thunk(
    cons(1, rot(nil, cons(2, nil), cons(3, nil)))
)
Back: nil
```

If we get the front of the queue or if we dequeue it, we will evaluate the thunk, which forces the evaluation of the rotation. Because the rotation is on an empty front list, it doesn't produce a new recursive rotation but directly produces a cons call:

```
q < front(q)
```

```
Front: lazy_thunk(
    cons(1, cons(2, cons(3, nil))))
)
Back: nil
```

At this point, the entire back list has been reversed and appended to the front queue, and from here on we can just dequeue the elements. The first dequeuing gets rid of the thunk as well as the first element:

```
q < dequeue(q)
```

```
Front: cons(2, cons(3, nil))
Back: nil
```

Other dequeue operations are now straightforward.

Let us go back to the empty list and insert six elements:

```
q <- empty_lazy_queue()
for (x in 1:6)
  q <- enqueue(q, x)
```

```
Front: lazy_thunk(
    cons(1, rot(nil, cons(2, nil), cons(3, nil))))
)
Back: cons(6, cons(5, cons(4, nil)))
```

Now, the front and back lists have the same length, so at the next enqueue operation, we will have to insert a rotation. This will force an evaluation of the thunk that is the front list, forcing, in turn, an evaluation of the rot function as we saw before, and then it will construct a new thunk wrapping this:

```
q <- enqueue(q, 7)

Front: lazy_thunk(
    cons(1, rot(cons(2, cons(3, nil)),
                    cons(6, cons(5, cons(4, nil))),
                    cons(7, nil)))
)
Back: nil
```

Accessing the front of the queue will trigger an evaluation of the thunk, calling one level of recursion on rot:

```
front(q)

Front: lazy_thunk(
    cons(1, cons(2, rot(cons(3, nil)),
                        cons(5, cons(4, nil)),
                        cons(6, cons(7, nil)))))
)
Back: nil
```

Dequeuing doesn't add another rotation because it wrapped in a thunk—although it is not so clear from my notation here, there is a thunk wrapping the rot call at the second level:

```
q <- dequeue(q)

Front: lazy_thunk(
    cons(2, rot(cons(3, nil)),
                cons(5, cons(4, nil)),
                cons(6, cons(7, nil))))
)
Back: nil
```

Calling front on this queue will again force an evaluation of rot:

```
front(q)
Front: lazy_thunk(
    cons(2, cons(3, rot(nil,
                        cons(4, nil),
                        cons(5, cons(6, cons(7, nil))))
)
Back: nil
```

You can continue the example, and I advise you to do that if you really want to understand how we simulate a data structure by modifying lazy expressions.

For the amortized complexity analysis, we can reason as before: whenever we add an element to the back queue, it also pays for moving the element to the front queue at a later time. So any sequence of operations will be bounded by the number of constant-time insertions, just as before. Because the lazy evaluation mechanism we have implemented remembers the result of evaluating an expression, we also get the same complexity if we use the queue persistently. If we need to perform expensive operations, which we will need when we remove elements from the front of the queue, this might be costly the *first* time we remove an element from a given queue. If we remove it again, because we do operations on a saved queue after we have modified it somewhere else, then the operation will be cheap.

For the worst-case complexity analysis, we first notice that enqueue operations always take constant time. We first add an element to the front of the back list, which is a constant time operation, and after that, we might call rot. Calling rot only constructs a thunk, though, which again is a constant time operation. We don't pay for rotations until we access the list a rot call wraps.

With `front` and `dequeue` we access lazy lists, so here we might have to actually call the rotation operation. The operation looks like it would be a constant time operation:

```
lazy_thunk(cons(car(front),
                rot(cdr(front), cdr(back), cons(car(back), a))))
```

This is a deception, however. We construct a new list from the first element in `front` and then add a thunk to the end of it, and this is a constant time operation if `front` is a simple list, but it *could* involve another call to `rot` that we need to call recursively. But if `front` is another call to `rot`, the front of the list we have in that call cannot be longer than half the length of `front` because we only construct `rot` thunks when the front is the same length as the back list. So we might have to call `rot` several times, but each time, the front list will have half the length as the previous. This means that we can at most have $\log(n)$ calls to `rot`, so each call to `front` and `dequeue` can occur at most involve logarithmically many operations.

To get constant worst-time operations, we need to do a little more rotation work in enqueuing operations, so these will pay for reversals. But before we get to that, I want to take a closer look at the `rot` function and notice some R specific technicalities.

In the `rot` function we `force` all the parameters:

```
rot <- function(front, back, a) {
  force(front)
  force(back)
  force(a)
  if (is_nil(front)) cons(car(back), a)
  else {
    lazy_thunk <- function(lst) function() lst()
```

```
    lazy_thunk(cons(car(front),
                    rot(cdr(front), cdr(back), cons(car(back),
                    a))))
  }
}
```

We typically have to force arguments to avoid problems that occur when expressions refer to variables that might have changed in the calling environment. That never happens in this queue implementation. Our queue implementation is purely functional, and we never modify any variables—in theory, we shouldn't have to force the parameters. It turns out that we *do* need to force them; at least we do need to force the accumulator, and exploring why gives us some insights into lazy evaluation in R.

Try modifying the function to look like this:

```
rot <- function(front, back, a) {
  if (is_nil(front)) cons(car(back), a)
  else {
    lazy_thunk <- function(lst) function() lst()
    lazy_thunk(cons(car(front),
                    rot(cdr(front), cdr(back), cons(car(back),
                    a))))
  }
}
```

If you then run this code

```
q <- empty_lazy_queue()
for (x in 1:10000) {
  q <- enqueue(q, x)
}
for (i in 1:10000) {
  q <- dequeue(q)
}
```

you will find that you call functions too deeply.[3]

 With the analysis we have done on how deep we will recurse when we rotated, that shouldn't happen. We might get a few tens deep in function calls with a sequence of ten thousand operations, but that shouldn't be a problem at all. What is going wrong?

 The problem is the lazy evaluation of the accumulator. In the recursive calls, we update the accumulator to be cons(car(back), a)), which is something we should be able to evaluate in constant time, but we don't actually do this. Instead, we pass the *expression* for the cons call along in the recursive call. It doesn't actually get evaluated until we access the accumulator. At *that* time, all the cons calls need to be evaluated, and that can involve a *lot* of function calls; we have to call cons for as many times as a is long. By not forcing a, we are delaying too much of the evaluation.

 The most common pitfall with R's lazy evaluation of parameters is the issue with changing variables. That problem might cause your functions to work incorrectly. This is a different pitfall that will give you the right answer if you get one but might involve evaluating many more functions than you intended. You avoid it by always being careful to force parameters when you return a closure, but in this case, we could also explicitly evaluate the cons(car(back),a) expression:

```
rot <- function(front, back, a) {
  if (is_nil(front)) cons(car(back), a)
  else {
    lazy_thunk <- function(lst) function() lst()
    tail <- cons(car(back), a)
    lazy_thunk(cons(car(front), rot(cdr(front), cdr(back), tail)))
  }
}
```

---

[3]How deep you can recurse depends on your R setup, so you might have to increase the number of elements you run through in the loops, but with a standard configuration this should be enough to get an error.

We are swimming in shark-infested waters when we use lazy evaluation, so we have to be careful. If we always `force` parameters when we can, though, we will avoid most problems.

Constant Time Lazy Queues

The rotation function performs a little of the reversal as part of constant time insertion operations, which is why we get a worst-case performance better than linear time. With making do just a little more, we can get constant time worst-case behavior. There will still be a slight overhead with this version compared to the ephemeral queue implementation, as illustrated in Figure 4-5, so if you don't need your queues to be persistent or fast for *every* operation, the first solution we had is still superior. If you *do* need to use the queue as a persistent structure, or you need to be guaranteed constant time operations, then this new variation is the way to go.

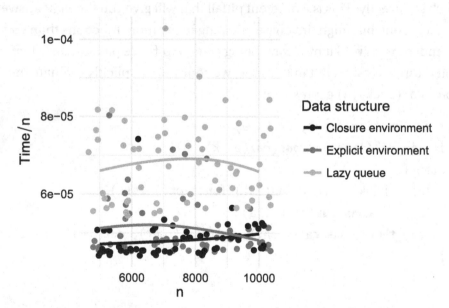

Figure 4-5. *Comparison of the ephemeral queue and the worst-case constant time lazy queue*

The worst-case constant time lazy queue uses the same rotation function as the $\log(n)$ worst-case queue, but it uses a helper list that is responsible for evaluating part of the front queue as part of other queue operations. With the rotation function, we handle the combination of concatenation and reversal to move elements from the back list to the front list as part of other operations. We will use the helper list to evaluate the thunks that the front list consist of as part of other operations. If we evaluate them as part of other operations, then, when we need to access the front list, the lazy expressions have been evaluated, and we can access the list in constant time.

Instead of keeping track of the length of the lists in the lazy_queue data structure, we just keep the extra helper list. We do not need to know the length of the lists in this implementation; we only need to check for empty lists:

```
lazy_queue <- function(front, back, helper) {
  structure(list(front = front, back = back, helper = helper),
            class = "lazy_queue")
}
```

We then modify the make_q function to this:

```
make_q <- function(front, back, helper) {
  if (is_nil(helper)) {
    helper <- rot(front, back, nil)
    lazy_queue(helper, nil, helper)
  } else {
    lazy_queue(front, back, cdr(helper))
  }
}
```

and we update the constructor and generic functions to reflect the changed data structure:

```
empty_lazy_queue <- function()
  lazy_queue(nil, nil, nil)
is_empty.lazy_queue <- function(x)
  is_nil(x$front) && is_nil(x$back)

enqueue.lazy_queue <- function(x, elm)
  make_q(x$front, cons(elm, x$back), x$helper)
front.lazy_queue <- function(x) car(x$front)
dequeue.lazy_queue <- function(x)
  make_q(cdr(x$front), x$back, x$helper)
```

You should think of helper more as a pointer into the front list than as a separate list itself. Its purpose is to walk through the elements in the front list and evaluate them, leaving just a list behind. The lazy lists we are working with are conceptually immutable, but because of lazy evaluation, some consist of simple cons calls and others of rot calls. As we progress through the helper function, we translate the recursive rot calls into cons calls— conceptually, at least; the actual list still consists of the rot call, but the value of the function has been evaluated. Because helper refers to a suffix of the front list at any time in the evaluation of queue operations, the effect of evaluating expressions in helper makes operations on front cheap.

Whenever we construct a queue with make_q, we either set up a new rotation of the front and back queue or we evaluate the first function in the helper list. We can see this in action through an example. We start with an empty list and insert an element. This involves a call to rot as in the earlier version of the lazy queue, but this time we also set the helper list to point to the front list:

```
q <- empty_lazy_queue()
q <- enqueue(q, 1)
```

```
Front: cons(1, nil)
Helper: cons(1, nil)
Back: nil
```

When we insert a second element, this gets prepended to the back list, and we make one step with the helper function, moving it to the empty list:

```
q <- enqueue(q, 2)
```

```
Front: cons(1, nil)
Helper: nil
Back: cons(2, nil)
```

With the third element, we again need a rotate call, since now the helper list is empty:

```
q <- enqueue(q, 3)
```

```
Front: cons(1, rot(nil, cons(2, nil), cons(3, nil)))
Helper: cons(1, rot(nil, cons(2, nil), cons(3, nil)))
Back: nil
```

If we now continue inserting elements into the queue, we can see how the helper function will walk through the front queue and evaluate the expressions there, updating the front queue at the same time as we insert new elements to the back queue:

```
q <- enqueue(q, 4)
```

```
Front: cons(1, cons(2, cons(3, nil)))
Helper: cons(2, cons(3, nil))
Back: cons(4, nil)
q <- enqueue(q, 5)
```

```
Front: cons(1, cons(2, cons(3, nil)))
Helper: cons(3, nil)
```

```
Back: cons(5, cons(4, nil))
q <- enqueue(q, 6)

Front: cons(1, cons(2, cons(3, nil)))
Helper: nil
Back: cons(6, cons(5, cons(4, nil)))
```

When we insert the seventh element, we again need a rotation, but now all the lazy expressions in the front queue have been evaluated:

```
q <- enqueue(q, 7)

Front: cons(1, rot(cons(2, cons(3, nil)),
                    cons(6, cons(5, cons(4, nil))),
                    cons(7, nil)))
Helper: cons(1, rot(cons(2, cons(3, nil)),
                    cons(6, cons(5, cons(4, nil))),
                    cons(7, nil)))
Back: nil
```

In future operations, helper will walk through the front queue again and evaluate tails of the function:

```
q <- enqueue(q, 8)

Front: cons(1, cons(2, rot(cons(3, nil),
                           cons(5, cons(4, nil)),
                           cons(6, cons(7, nil)))))
Helper: cons(2, rot(cons(3, nil),
                    cons(5, cons(4, nil)),
                    cons(6, cons(7, nil))))
Back: cons(8, nil)
q <- enqueue(q, 9)
```

```
Front: cons(1, cons(2, cons(3,
          rot(nil,
              cons(4, nil),
              cons(5, cons(6, cons(7, nil))))))))
Helper: cons(3, rot(nil,
                    cons(4, nil),
                    cons(5, cons(6, cons(7, nil))))))
Back: cons(9, cons(8, nil))
q <- enqueue(q, 10)

Front: cons(1, cons(2, cons(3,
          cons(4, cons(5, cons(6, cons(7, nil))))))))
Helper: cons(4, cons(5, cons(6, cons(7, nil))))
Back: cons(10, cons(9, cons(8, nil)))
```

We don't just progress the evaluation of the helper function when we add elements; we also move it forward when removing elements:

```
q <- dequeue(q)

Front: cons(2, cons(3,
          cons(4, cons(5, cons(6, cons(7, nil)))))))
Helper: cons(5, cons(6, cons(7, nil)))
Back: cons(10, cons(9, cons(8, nil)))
q <- dequeue(q)

Front: cons(3, cons(4, cons(5, cons(6, cons(7, nil)))))
Helper: cons(6, cons(7, nil))
Back: cons(10, cons(9, cons(8, nil)))
q <- dequeue(q)

Front: cons(4, cons(5, cons(6, cons(7, nil))))
Helper: cons(7, nil)
Back: cons(10, cons(9, cons(8, nil)))
```

```
q <- enqueue(q, 11)
```

```
Front: cons(4, cons(5, cons(6, cons(7, nil))))
Helper: nil
Back: cons(11, cons(10, cons(9, cons(8, nil))))
```

We do not insert the next rotation until the helper list is empty, at which point the entire front list has been lazily evaluated. It is exactly because of this, that the helper function goes through the front list and performs rotations one element in the list at a time, that the worst-case performance is constant time.

With a little more work, it is possible to adapt this strategy to double-ended queues as well. For that construction, I refer to Okasaki (1995b).

Explicit Rebuilding Queue

The lazy queue with constant time operations works by moving elements from the back list to the front list a little at a time as part of other operations. This idea can also be implemented without lazy evaluation by explicitly keeping track of some extra lists that capture the intermediate steps in moving elements.

We will rebuild the front queue as part of the enqueue and dequeue operations by having an extra copy of it in an intermediate state in two or four other lists. We can split the movement of elements from the back list to the front queue into two phases. We first reverse both lists—the back list because we need its element in reverse order when they are put in the front list, and the front list because we can prepend the elements in this list to the back list one at a time if we have reversed it first. When we reverse the lists, we will represent them as four lists, the elements we haven't reversed in their original order in a front and back list, and the elements we have reversed in a reversed_front and reversed_back list. One reversal operation will consist of moving one element from front to reverse_front and one element from back to reversed_back. When we

124

are done with the reversal, we represent the partial appending as two lists, front and back, and in each operation, we move one element from back to front. To move n elements from the back list to the front list via these two operations, reverse and append, it takes $2n$ operations—n reversal operations and n appending operations. If we start this rebuilding of the front list whenever the back list has the same length as the front list, and we do two operations per enqueue and dequeue, we are guaranteed to have built a new front list before we have emptied the existing one (we will have emptied it if we perform n dequeue operations). Using this strategy, we can build an updated front list by making two extra operations each time we modify the queue.

To keep track of where we are in the rebuilding of the front list, we define four states:

- IDLE that we use when we shouldn't be doing any update because we are not actively moving any elements

- REVERSING when we are reversing the lists,

- APPENDING when we are appending lists

- DONE when we have finished building a new front list and are ready to update the queue with it

When we are in the REVERSING state, we keep track of four extra lists, as described, and when we are in the APPENDING state, we keep track of two extra lists. When we are in the IDLE state, we do not need to keep track of any extra lists, and when we are in the DONE state, we need to remember the result of all the operations until we actually update the queue with the new front list. We define these structures to keep track of the states:

```
IDLE <- 0
REVERSING <- 1
APPENDING <- 2
DONE <- 3
```

```
idle_state <- function() list(state = IDLE)
reversing_state <- function(
  keep
  , front
  , reverse_front
  , back
  , reverse_back) {
  list(state = REVERSING,
       keep = keep,
       front = front,
       reverse_front = reverse_front,
       back = back,
       reverse_back = reverse_back)
}
appending_state <- function(
  keep
  , front
  , back) {
  list(state = APPENDING,
       keep = keep,
       front = front,
       back = back)
}
done_state <- function(result) {
  list(state = DONE, result = result)
}
```

Ignore the keep variable for now; it is one we need to keep track of elements that we *should not* append to the new front list because they have been dequeue'ed while we were rebuilding the list. It will keep track of how many elements we should move from the reversed front list to the reversed back list. When we add elements to the reversed front list, we increase it, and when we move elements to the back list in append operations, we decrease it. If we perform a dequeue operation we also decrease it, so we prevent the dequeued element from being added to the updated front list—see the following invalidate function.

The data structure has the front and back lists and keeps track of their sizes, so we know when to start moving elements into the new front list, and it contains the state, so we know what our next step and our intermediate lists are. We define it like this:

```
rebuild_queue_node <- function(
  state
  , front_size
  , front
  , back_size
  , back) {
  structure(list(state = state,
                 front_size = front_size,
                 front = front,
                 back_size = back_size,
                 back = back),
          class = "rebuild_queue")
}
```

Constructing an empty queue and testing for emptiness is straightforward. Because we always start moving element from back to front in time to have the movements completed before front is empty (unless the entire queue is empty), we can always get the front element from the front list, so the front operation is also simple to implement:

```
empty_rebuild_queue <- function() {
  rebuild_queue_node(state = idle_state(),
                     front_size = 0,
                     front = empty_list(),
```

```
                         back_size = 0,
                         back = empty_list())
}

is_empty.rebuild_queue <- function(x) is_empty(x$front)
front.rebuild_queue <- function(x) {
  list_head(x$front)
}
```

When performing an update operation, we check which state we are in and perform the appropriate operation based on that. For the REVERSING state we should move an element from the front list to the reverse_front list and an element from the back list to the reverse_back list. If the front list is empty, we are done with reversing and should start appending. When we start reversing, it is because the back list in the queue has gotten one longer than the front list, so when we finish reversing, we have to remember to move one element from the back to reverse_back list that we will start appending. When we are appending, we need to move one element from front to back in each step, and we are done when the keep counter is zero. We need to use the keep counter instead of checking if front is empty, because if we have performed any dequeue operations, then the front list in the APPENDING state contains elements that *shouldn't* be added to the updated front list:

```
exec <- function(state) {
  if (state$state == REVERSING) {
    if (is_empty(state$front)) {
      appending_state(keep = state$keep,
                      front = state$reverse_front,
                      back = list_cons(
                        list_head(state$back),
                        state$reverse_back)
                      )
```

```
    } else {
    reversing_state(keep = state$keep + 1,
                    front = list_tail(state$front),
                    reverse_front = list_cons(
                        list_head(state$front),
                        state$reverse_front),
                    back = list_tail(state$back),
                    reverse_back = list_cons(
                        list_head(state$back),
                        state$reverse_back))
    }
  } else if (state$state == APPENDING) {
    if (state$keep == 0) {
      done_state(result = state$back)
    } else {
      appending_state(keep = state$keep - 1,
                      front = list_tail(state$front),
                      back = list_cons(
                        list_head(state$front),
                        state$back))
    }
  } else {
    state
  }
}
```

For each operation that updates the queue, we need to perform two update operations, so we wrap that in the following function. Here we also check whether we have reached the DONE state after performing the two

operations, in which case we replace the front queue with the result list
we store in this state and set the new state to IDLE:

```
exec2 <- function(x) {
  new_state <- exec(exec(x$state))
  if (new_state$state == DONE)
    rebuild_queue_node(state = idle_state(),
                       front_size = x$front_size,
                       front = new_state$result,
                       back_size = x$back_size,
                       back = x$back)
  else
    rebuild_queue_node(state = new_state,
                       front_size = x$front_size,
                       front = x$front,
                       back_size = x$back_size,
                       back = x$back)
}
```

To set up a new rebuilding of the front list when the back list becomes
longer than the front list, we wrap updates of the queue in calls to a check
function. It will call exec2 as long as the back list is shorter than or equal in
length to the front list; otherwise it will set the state to REVERSAL. When it
does this, the state front list is set to the current queue front list, and the
state back list is set to the queue back list, and the two reverse lists are set
to empty lists. The keep counter is set to zero. The new queue we construct
with this state has an empty back list, and we set the length of the front list
to the length of the front list once we have updated it. The actual front list
will not have this length yet, but we pretend that it does for future checks of

the lists lengths, and conceptually it does have this length—we are just not quite done with constructing the list yet:

```r
check <- function(x) {
  if (x$back_size <= x$front_size) {
    exec2(x)
  } else {
    # when back gets longer than front, we start reversing
    new_state <- reversing_state(keep = 0,
                                 front = x$front,
                                 reverse_front = empty_list(),
                                 back = x$back,
                                 reverse_back = empty_list())
    new_queue <- rebuild_queue_node(state = new_state,
                                    front_size = x$front_size +
                                    x$back_size,
                                    front = x$front,
                                    back_size = 0,
                                    back = empty_list())
    exec2(new_queue)
  }
}
```

Enqueueing elements is not just a simple matter of adding an element to the back list and calling check:

```r
enqueue.rebuild_queue <- function(x, elm) {
  check(rebuild_queue_node(state = x$state,
                           front_size = x$front_size,
                           front = x$front,
                           back_size = x$back_size + 1,
                           back = list_cons(elm, x$back)))
}
```

Dequeuing is slightly more complicated. We need to remove the front element from both the list we actually store in the queue and the list we are in the process of rebuilding. For the latter, we use the keep counter. When we dequeue, we invalidate the current state by calling the following function `invalidate`. This function decreases the keep counter in the current state if we are in the process of rebuilding a front list. It is possible that the last update of the state was an appending operation that decreased keep to zero. If this is the case, we need to set the state to DONE instead and remove the first element of the result we have built:

```r
invalidate <- function(state) {
  if (state$state == REVERSING) {
    reversing_state(keep = state$keep - 1,
                    front = state$front,
                    reverse_front = state$reverse_front,
                    back = state$back,
                    reverse_back = state$reverse_back)
  } else if (state$state == APPENDING) {
    if (state$keep == 0) {
      done_state(result = list_tail(state$back))
    } else {
      appending_state(keep = state$keep - 1,
                      front = state$front,
                      back = state$back)
    }
  } else {
      state
  }
}
```

Dequeuing is now simply a question of invalidating the state and removing the front element in the queue's front list as well, and then we check the result to make sure that we perform two update operations as part of the dequeue function:

```
dequeue.rebuild_queue <- function(x) {
  new_queue <- rebuild_queue_node(state = invalidate(x$state),
                                  front_size = x$front_size - 1,
                                  front = list_tail(x$front),
                                  back_size = x$back_size,
                                  back = x$back)
    check(new_queue)
}
```

This rebuilding queue is mostly of theoretical interest. When it comes to its actual performance, the operations we perform impose a large overhead that makes this queue implementation slower than the other we have considered, as shown in Figure 4-6.

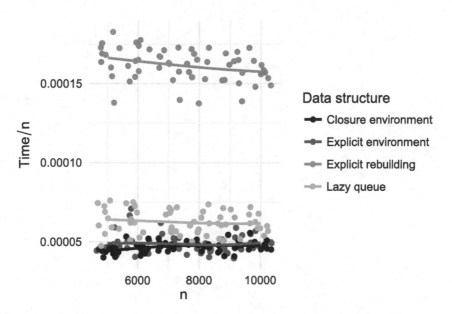

Figure 4-6. *Comparison of queue performance for a series of enqueue and dequeue operations*

CHAPTER 5

Heaps

Heaps, or priority queues, are collections of elements from an ordered set where besides checking for emptiness and inserting elements

```
is_empty <- function(x) UseMethod("is_empty")
insert <- function(x, elm) UseMethod("insert")
```

we can also access and delete the smallest element:[1]

```
find_minimal <- function(heap) UseMethod("find_minimal")
delete_minimal <- function(heap) UseMethod("delete_minimal")
```

Heaps do not necessarily contain sets. It is possible for heaps to have multiple elements with the same value.

In addition to these operations, we will also require that we can merge two heaps:

```
merge <- function(x, y) UseMethod("merge")
```

In many of the implementations, the merge function is useful, and we can always implement a default version like this:

```
merge.default <- function(x, y) {
  while (!is_empty(y)) {
    x <- insert(x, find_minimal(y))
```

---

[1]I will implement all the heaps in this chapter to have access to the minimal element. It is a trivial modification to have access to the largest instead.

© Thomas Mailund 2017
T. Mailund, *Functional Data Structures in R*, https://doi.org/10.1007/978-1-4842-3144-9_5

```
    y <- delete_minimal(y)
  }
  x
}
```

Because we already used merge with bags, though, we probably shouldn't make this default implementation of the generic function. Instead, we can make a version for heaps

```
merge.heap <- function(x, y) {
  while (!is_empty(y)) {
    x <- insert(x, find_minimal(y))
    y <- delete_minimal(y)
  }
  x
}
```

and make all our heap implementations inherit from a generic "heap" class.

This heap merging will work for all heaps, but usually there are more efficient solutions for merging heaps. This default solution will require a number of find_minimal and delete_minimal operations equal to the size of heap y, and at the very least this will be linear in the size of y. As we will see, we can usually do better.

One use of heaps is for sorting elements. This approach is known as *heap sort*. If we have a heap, we can construct a list from its element in reverse order using a loop similar to the merge.default function:

```
heap_to_list <- function(x) {
  l <- empty_list()
  while (!is_empty(x)) {
    l <- list_cons(find_minimal(x), l)
    x <- delete_minimal(x)
```

```
    }
    l
}
```

This function creates a linked list from a heap, but the elements are added in decreasing rather than increasing order, so to implement a full sort we need to reverse it. Of course, to sort a vector of elements, we also need to construct the heap from the elements. One general approach is to just insert all the elements into a heap, starting from the empty heap:

```
vector_to_heap <- function(empty_heap, vec) {
  heap <- empty_heap
  for (e in vec)
    heap <- insert(heap, e)
  heap
}
```

With the vector_to_heap and heap_to_list functions, we can sort elements like this:

```
heap_sort <- function(vec, empty_heap) {
  heap <- vector_to_heap(empty_heap, vec)
  lst <- heap_to_list(heap)
  list_reverse(lst)
}
```

The time complexity of the heap_sort function depends on the time complexity of the heap operations. We can reverse lst in linear time. If insert, find_minimal, and delete_minimal all run in logarithmic time, the entire sort can be done in time $O(n\log(n))$, which is optimal for comparison-based sorting. Because we know that this is an optimal time for sorting, it also gives us some hints at how efficient we can hope the heap data structures to be. Either vector_to_heap or heap_to_list *must* take time $O(n \log(n))$ as a tight upper bound on the complexity if we have an optimal sorting algorithm.

As it turns out, we can usually construct a heap from n elements in linear time, but it will take us $O(n \log(n))$ time to extract the elements again. Constructing heaps in linear time uses the merge function. Instead of adding one element at a time, we construct a sequence of heaps and merge them. The construction looks like this:

```
singleton_heap <- function(empty_heap, e) insert(empty_heap, e)
vector_to_heap <- function(vec, empty_heap, empty_queue) {
  q <- empty_queue
  for (e in vec)
    q <- enqueue(q, singleton_heap(empty_heap, e))
  repeat {
    first <- front(q) ; q <- dequeue(q)
    if (is_empty(q)) break
    second <- front(q) ; q <- dequeue(q)
    new_heap <- merge(first, second)
    q <- enqueue(q, new_heap)
  }
  first
}
```

We start by constructing a singleton heap containing just one of the input elements. Then we put these in a queue, and while we have more than one heap, we take the first two out of the queue, merge them, and put them back at the end of the queue. You can think of this as going through several *phases* of heaps. The first phase merges heaps of size one into heaps of size two. The second phase merges heaps of size two into heaps of size four. And so on. Because we can only double the heap size logarithmically many times, we have $\log(n)$ phases.

The size of the heaps we merge in phase m is 2^m, but if we can merge heaps in logarithmic time, we get the total time for constructing the heap in this way to be

$$\sum_{m=0}^{\log(n)} \frac{n}{2^m} O(m) = O\left(n \sum_{m=0}^{\log(n)} \frac{m}{2^m} \right)$$

which, because the series

$$\sum_{m=0}^{\infty} \frac{m}{2^m}$$

is convergent, is $O(n)$. So, if we can merge heaps in logarithmic time, we can construct a heap of n elements in time $O(n)$.[2] This also tells us that if we can merge heaps in logarithmic time, it must also cost us at least logarithmic time to either get the minimal value or delete it—otherwise we would be able to sort faster than $O(n\log(n))$, which we know is impossible for comparison-based sorting.

With this implementation of heap construction, we would have to modify the `heap_sort` function to look like this:

```
heap_sort <- function(vec, empty_heap, empty_queue) {
  heap <- vector_to_heap(vec, empty_heap, empty_queue)
  lst <- heap_to_list(heap)
  list_reverse(lst)
}
```

The function is parameterized with the empty heap and queue, which lets us choose which data structures to use. We know that the fastest queue

---

[2] I am being a little lax with notation here. We cannot sum to $\log(n)$ unless this is an integer, and strictly speaking it should be $\lfloor \log(n) \rfloor$. This doesn't change the complexity analysis and I just don't feel like being too strict with the mathematics in a book that is primarily about R programming.

implementation we have, when we do not need the queue to be persistent, is the environment-based queue, so we could put that as the default empty queue to get a good algorithm. In the remainder of this chapter, we will experiment with different heap implementations to see which heap would be a good default choice.

The heaps we will implement are based on trees, so they are slightly more complicated than the lists and queues we used in the previous chapter—but only slightly so. When we work with trees, we say that a tree has the "heap property" if every node in the tree contains a value and the nodes in the trees satisfy that all children of a node have values greater than the value in the node. By recursive reasoning, this means that if we have a tree with the heap property, then all subtrees also have the heap property. This will be an invariant that we will ensure for all the trees we work with.

Leftist Heaps

Leftist heaps are a classical implementation of heaps based on one simple idea: you represent the heap as a binary tree with the heap property, and you make sure that the left subtree is always at least as large as the right subtree.[3] The structure exploits a general trick in data structure design known as the *smaller half trick*. If you have to do some computation recursively, but you only have to do it for one of two children in a binary tree, you pick the smaller of the two trees. If you do this, you slice away half of the size of the tree in each recursive call.[4] If you slice away half the data

---

[3]My description of leftist heaps is based on Okasaki (1999a). The original description of the data structure can be found in Crane (1972).

[4]The *smaller half* name is an oxymoron. If we really split data in half, both halves would be the same size. What we really mean is that we slice away at least one half of the data in each recursion. I didn't come up with the name, but that is the name I know the trick by.

in each recursive call, you can never recurse deeper than the logarithm of the full data size.

In the leftist heap, we don't actually keep track of the full size of heaps to exploit this trick. We don't need trees to be balanced to operate efficiently on heaps. Instead, we worry about the total depth we might have to go to in recursions. We will always recurse to the right in the heap, so the length of the rightmost path in a heap, which we will call its *rank*, is what we worry about, and we make sure that we always have the shortest path as the rightmost.

A variant of this called *maxiphopic heaps* (Okasaki 2005) makes this more explicit, but the underlying idea is the same: maxiphopic heaps just do not require that the larger subtree be the left tree.

To keep the invariant that the left subtree always have a smaller rank than the right subtree, we keep track of tree ranks. The structure we use to represent a leftist heap looks like this:

```
leftist_heap_node <- function(
  value
  , left = empty_leftist_heap()
  , right = empty_leftist_heap()
  , rank = 0
  ) {
  structure(list(left = left,
                 value = value,
                 right = right,
                 rank = rank),
            class = c("leftist_heap", "heap"))
}
```

The class of the structure is "leftist_heap" and then "heap", in case we have general heap functions that are not specific to leftist heaps.

The empty list and the emptiness test uses the sentinel trick as we have used it earlier. We do not test identity against the sentinel for emptiness

tests, though, to avoid the issues this could give us if we would at some
point modify a leaf (as we will when we implement plotting code later):

```
empty_leftist_heap_node <- leftist_heap_node(NA, NULL, NULL)
empty_leftist_heap <- function() empty_leftist_heap_node
is_empty.leftist_heap <- function(x)
  is.null(x$left) && is.null(x$right)
```

Because a leftist heap has the heap property, we can always get the
minimum value from the root of the tree:

```
find_minimal.leftist_heap <- function(heap) {
  heap$value
}
```

To delete the minimum value, we need to get rid of the value in the
root. But because the two subtrees of the root are heaps, we can create
a new heap with the minimal value removed just by merging the two
subtrees:

```
delete_minimal.leftist_heap <- function(heap) {
  merge(heap$left, heap$right)
}
```

Inserting an element is equally simple: we can make a singleton heap
and merge it into the existing heap:

```
insert.leftist_heap <- function(x, elm) {
  merge(x, leftist_heap_node(elm))
}
```

All the complexity of a leftist heap boils down to the merge operation.
This is where we will exploit the fact that the right subtree is never more
than half as deep as the full heap. If we merge two heaps, we take the
minimal value in the root, put the left part of the first tree as the left

subtree, and then merge recursively on the right. Because this cuts the problem down to half the size, we will never spend more time than $O(\log n)$:

```
build_leftist_heap <- function(value, a, b) {
  if (a$rank >= b$rank)
    leftist_heap_node(value = value,
                      left = a,
                      right = b,
                      rank = b$rank + 1)
  else
    leftist_heap_node(value = value,
                      left = b,
                      right = a,
                      rank = a$rank + 1)
}

merge.leftist_heap <- function(x, y) {
  if (is_empty(x)) return(y)
  if (is_empty(y)) return(x)
  if (x$value <= y$value)
      build_leftist_heap(x$value, x$left, merge(x$right, y))
  else
      build_leftist_heap(y$value, y$left, merge(x, y$right))
}
```

The base cases of the merge operation ensures that if either of the two heaps we merge is empty, we return the other. Otherwise, we call recursively on the right. We always put the smallest value in the root, to preserve the heap property, but after that we call recursively on half the problem. The smaller half trick does the trick for us to get logarithmic complexity.

Because we can immediately get the smallest value from the root of a leftist heap, the `find_minimal` operation runs in $O(1)$. The `insert` and `delete_minimal` operations work through the `merge` operation, so these operations run in time $O(\log n)$.

I hope you agree that this implementation of a heap is straightforward. There is only one trick to it—the smaller half trick—and the rest is just remembering to call recursively to the right and never to the left. It is surprising how powerful the smaller half trick really is. I have used it in many different algorithms, and I am always awed by how powerful that simple trick is.

But just because we have one solution to the heap data structure there is no reason to stop. It is worth exploring other solutions. Even if they are less efficient, which they might be if they are more complex to implement, there may be some insights to gain from implementing them.

Binomial Heaps

For the second version of heaps, we turn to *binomial heaps*. My presentation of this data structure is also based on Okasaki (1999a). Binomial heaps are based on *binomial trees*, which are trees with the heap structure and the additional invariants (see Figure 5-1):

- A binomial tree of rank 0 is a singleton.

- A tree of rank r has r children, $t_1, t_2, ..., t_r$ where t_i is a binomial tree with rank $r - i$.

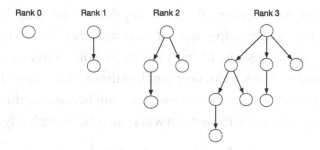

Figure 5-1. Binomial trees

We are not going to use binomial trees as an abstract data structure, so we won't give them a class and just implement them using a `list`:

```
binomial_tree_node <- function(value, trees) {
  list(value = value, trees = trees)
}
```

We will build up binomial trees by *linking* them. This is an operation that we will only do on trees with the same rank, and what the operation does is make one of the trees the leftmost subtree of the other. If both trees have rank r, then this constructs a new tree of rank $r + 1$. To preserve the heap property, we must make sure that the parent tree is the one with the smallest value of the two. We can implement this operation as such:

```
link_binomial_trees <- function(t1, t2) {
  if (t1$value < t2$value) {
    binomial_tree_node(t1$value, list_cons(t2, t1$trees))
  } else {
    binomial_tree_node(t2$value, list_cons(t1, t2$trees))
  }
}
```

Binomial trees are not themselves an efficient approach to building heaps. In fact, we cannot use them as heaps at all. We can, of course, easily get the minimal value from the root, but we cannot represent an arbitrary

145

number of elements in binomial trees—they don't come in all sizes because of the invariants—and manipulation of binomial trees does not easily allow the heap operations. Instead, a binomial heap is a list of trees, each with their associated rank so we can keep track of those. The minimal value in a heap will be in one of the roots of these trees, but because finding it would require searching through the list, we will remember it explicitly:

```r
binomial_heap_node <- function(rank, tree) {
  list(rank = rank, tree = tree)
}
binomial_heap <- function(min_value, heap_nodes = empty_list())
{
  structure(list(min_value = min_value, heap_nodes = heap_nodes),
            class = c("binomial_heap", "heap"))
}
```

With this structure, an empty binomial heap is just one with no binomial trees, and we don't need a sentinel to represent such:

```r
empty_binomial_heap <- function() binomial_heap(NA)
is_empty.binomial_heap <- function(x) is_empty(x$heap_nodes)
```

Because we explicitly represent the minimal value in the heap, the find_minimal function is trivial to implement:

```r
find_minimal.binomial_heap <- function(heap) {
  heap$min_value
}
```

We now insist on the following invariant for how the binomial trees are used in a binomial heap: no two trees can have the same rank. This creates a correspondence between the rank of binomial trees in a heap and the binary representation of the number of elements in the heap: for each 1 in the binary representation, we will have a tree of that rank, as shown in Figure 5-2.

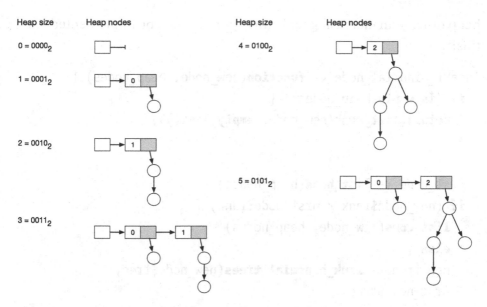

Figure 5-2. *Binomial heaps of size 0 to 5*

With this invariant in mind, we can think of both insertions and merge as a variant of binary addition. Insertion is the simplest case, so let's deal with that first. To insert a new value in a heap, we first create a singleton heap node, with a binomial tree of rank 0 holding the value:

```
singleton_binomial_heap_node <- function(value) {
  tree <- binomial_tree_node(value, empty_list())
  binomial_heap_node(0, tree)
}
```

We now need to insert this node in the list. If there is no node in there with rank 0 already, we can just put it in. If there is, though, that slot is taken, so we must do something else. We can link the existing tree of rank 0 with the new singleton, creating a node with rank 1. If that slot is free, we are done; if it is not, we must link again, and so on, similarly to how we carry a bit if we add binary numbers. If we always keep the trees in a heap ordered in increasing rank, this approach can be implemented like this:

```
insert_binomial_node <- function(new_node, heap_nodes) {
  if (is_empty(heap_nodes)) {
    return(list_cons(new_node, empty_list()))
  }

  first_node <- list_head(heap_nodes)
  if (new_node$rank < first_node$rank) {
    list_cons(new_node, heap_nodes)
  } else {
    new_tree <- link_binomial_trees(new_node$tree,
    first_node$tree)
    new_node <- binomial_heap_node(new_node$rank + 1, new_tree)
    insert_binomial_node(new_node, list_tail(heap_nodes))
  }
}
```

The insert operation on the heap now consists of updating the minimal value, if necessary, and inserting the new value starting from a singleton:

```
insert.binomial_heap <- function(x, elm, ...) {
  new_min_value <- min(x$min_value, elm, na.rm = TRUE)
  new_node <- singleton_binomial_heap_node(elm)
  new_nodes <- insert_binomial_node(new_node, x$heap_nodes)
  binomial_heap(new_min_value, new_nodes)
}
```

The na.rm = TRUE is necessary here to deal with the case where the heap is empty. We could have avoided it by using Inf as the value for an empty heap as well, but I find it nicer to explicitly state that an empty heap doesn't actually have a minimal value.

148

Merging two heaps also works similarly to binary addition. We have the two heaps represented as lists of binary trees in increasing rank order, so we can implement this as list merge. Whenever the front of one list has a rank smaller than the front of the other, we can insert that element in the front of a list and make a recursive call, but when the two lists have fronts of equal rank, we must link the two and merge the new tree in. We cannot simply put the new tree at the front of the merge because the existing lists might already have a slot for the rank of that tree, but we can insert it into the result of a recursive call, which will work like carrying a bit in addition:

```
merge_heap_nodes <- function(x, y) {
  if (is_empty(x)) return(y)
  if (is_empty(y)) return(x)

  first_x <- list_head(x)
  first_y <- list_head(y)
  if (first_x$rank < first_y$rank) {
    list_cons(first_x, merge_heap_nodes(list_tail(x), y))
  } else if (first_y$rank < first_x$rank) {
    list_cons(first_y, merge_heap_nodes(list_tail(y), x))
  } else {
    new_tree <- link_binomial_trees(first_x$tree, first_y$tree)
    new_node <- binomial_heap_node(first_x$rank + 1, new_tree)
    rest <- merge_heap_nodes(list_tail(x), list_tail(y))
    insert_binomial_node(new_node, rest)
  }
}
```

The actual merge operation just needs to keep track of the new minimal value in addition to merging the heap nodes:

```
merge.binomial_heap <- function(x, y, ...) {
  if (is_empty(x)) return(y)
  if (is_empty(y)) return(x)
  new_min_value <- min(x$min_value, y$min_value)
  new_nodes <- merge_heap_nodes(x$heap_nodes, y$heap_nodes)
  binomial_heap(new_min_value, new_nodes)
}
```

We don't need na.rm = TRUE in this case because we handle empty heaps explicitly.[5]

The insertion operation is really just a special case of the merge, as it was for leftist heaps, and we could have implemented it in terms of merging as this:

```
insert_binomial_node <- function(new_node, heap_nodes) {
  merge_heap_nodes(list_cons(new_node, empty_list()),
  heap_nodes)
}
```

Here, we just need to make the new node into a list and merge that into the existing heap nodes.

The complexity of the merge operation comes from the correspondence to binary numbers. To represent a number of size n we only need $\log n$ bits, so for heaps of size n the lists are no longer than $\log n$. We are not simply merging them but have the added complexity of having to carry a bit, but even doing this, which is simply binary addition, the complexity remains $O(\log n)$—the complexity of adding two numbers of

---

[5]If someone inserts *NA* into a heap, this would break, of course, but then if someone does that they should have his head examined. With *NA* there is no ordering, so the whole purpose of having a priority queue goes out the window.

size n represented in binary. Because insertion is just a special case of merging, we can, of course, insert in $O(\log n)$ as well.

Deleting the minimal value from a binomial heap is not really a more complex operation, it just involves a lot more code because we need to manipulate lists. The minimal value is found at the root of one of the trees in the heap. We need to find this tree, and we need to delete it from the list of trees. We could do this in one function returning two values, but that would involve wrapping and unwrapping the return value, so we will handle it in two operations instead. We know which value to search for in the roots from the saved minimal value in the heap, so finding the tree containing it is just a linear search through the heap nodes, and deleting it is just as simple:

```
get_minimal_node <- function(min_value, heap_nodes) {
  first_node <- list_head(heap_nodes)
  if (first_node$tree$value == min_value) first_node
  else get_minimal_node(min_value, list_tail(heap_nodes))
}

delete_minimal_node <- function(min_value, heap_nodes) {
  first_node <- list_head(heap_nodes)
  if (first_node$tree$value == min_value) {
    list_tail(heap_nodes)
  } else {
    rest <- delete_minimal_node(min_value, list_tail
    (heap_nodes))
    list_cons(first_node, rest)
  }
}
```

These are linear time operations in the length of the list, but because the list we operate on cannot be longer than $O(\log n)$, we can do them in logarithmic time in the size of the heap.

151

Deleting the tree containing the smallest value certainly gets rid of that value, but also any other values that might be in that tree. We need to put those back into the heap. We will do this by merging them into the heap nodes. To do that, though, we need to associate them with their rank; we need to wrap them in the heap node structure. If the tree we are deleting has rank r, then we know that its first subtree has rank $r - 1$, its second has rank $r - 2$, and so forth, so we can iterate through the trees, carrying the rank to give to the front tree along, in a recursion that looks like this:

```
binomial_trees_to_nodes <- function(rank, trees) {
  if (is_empty(trees)) {
    empty_list()
  } else {
    list_cons(binomial_heap_node(rank, list_head(trees)),
              binomial_trees_to_nodes(rank - 1, list_
              tail(trees)))
  }
}
```

If the tree we remove has rank r, this is an $O(r)$ operation. But because the ranks of the trees cannot be larger than $O(\log n)$—again, think of the binary representation of a number—this is a logarithmic operation in the size of the heap. We need to merge them into the original list, with the minimal tree removed, but they are in the wrong order. The children of a binomial tree are ordered in decreasing rank, but the trees in a binomial heap are ordered in increasing rank, so we will have to reverse the list before we merge. This we can also do in time $O(\log n)$ because it is a linear time operation in the length of the list.

In case you are wondering why we didn't just represent the children of binary trees in increasing order to begin with, it has to do with how we link two trees. We can link two trees in constant time because it just involves prepending a tree to a list of trees. If we wanted to store the trees in increasing rank order, we would need to append a tree to a list instead.

This would either require linear time in the size of the trees or a more complex data structure. It is easier just to reverse the list at this point.

Because we are deleting the minimal value of the heap, we also need to update the value we store for that. Here we can run through the new list once it is constructed and find the smallest value in the roots of the trees:

```
binomial_nodes_min_value <- function(heap_nodes, cur_min = NA)
{
  if (is_empty(heap_nodes)) {
    cur_min
  } else {
    front_value <- list_head(heap_nodes)$tree$value
    new_cur_min <- min(cur_min, front_value, na.rm = TRUE)
    binomial_nodes_min_value(list_tail(heap_nodes), new_cur_min)
  }
}
```

We drag the current minimal value along as an accumulator and give it the default value of NA. Therefore, we also need to use na.rm = TRUE when updating it, but using NA as its default value also guarantees that if we construct an empty heap when deleting the last element, it gets the minimal value set to NA.

All these operations take time $O(\log n)$, and the delete_minimal operation is just putting them all together, so we have a $O(\log n)$ operation that looks like this:

```
delete_minimal.binomial_heap <- function(heap) {
  min_node <-
    get_minimal_node(heap$min_value, heap$heap_nodes)
  other_nodes <-
    delete_minimal_node(heap$min_value, heap$heap_nodes)
  min_node_nodes <-
    binomial_trees_to_nodes(min_node$rank - 1,
                            min_node$tree$trees)
```

```
  new_nodes <-
    merge_heap_nodes(other_nodes, list_reverse(min_node_nodes))
  new_min_value <- binomial_nodes_min_value(new_nodes)
  binomial_heap(new_min_value, new_nodes)
}
```

In summary, we can implement a binomial heap with $O(1)$ find_minimal and $O(\log n)$ insert, merge and delete_minimal worst-case complexity. We can show, however, that insert actually runs in time $O(1)$ amortized by considering how the list of heap nodes behave compared to how many link operations we make. If you consider the original insert_binomial_node implementation, it is clear that we only recurse when we make a link operation, so the complexity of the function is the number of link operations plus one. You can think of each link operation as switching a one bit in the original heap list binary number to zero and the termination of the recurse as switching one zero bit to one. If we now think of switching a bit from zero to one as costing two credits instead of one, then such operations also pay for flipping them back to zero again in a later insertion. This analysis, however, is only valid if we consider the heap an ephemeral data structure—if we consider it a persistent data structure, nothing prevents us from spending the credits on the one bits more than once. The other $O(\log n)$ worst-case operations are still $O(\log n)$ when amortized.

Because of the amortized constant time insertion operation, constructing binomial heaps is much more efficient than constructing leftist heaps. Figure 5-3 shows experiments constructed using the following experiment code:

```
setup <- function(n) n
evaluate <- function(empty) function(n, x) {
  elements <- 1:n
  heap <- empty
```

```
for (elm in elements) {
  heap <- insert(heap, elm)
  }
}
```

When we divide the time it takes to construct a heap of a given size by that size, we see that the binomial heap has a flat time curve while the leftist heap grows logarithmically, telling us that in practice, the running time to construct a binomial heap is $O(n)$ while the time to construct a leftist heap is $O(n\log n)$. This justifies the much more complex data structure that characterize binomial heaps compared to leftist heaps, at least when we don't need a persistent data structure. The amortized analysis only works when we treat the binomial heap ephemerally.

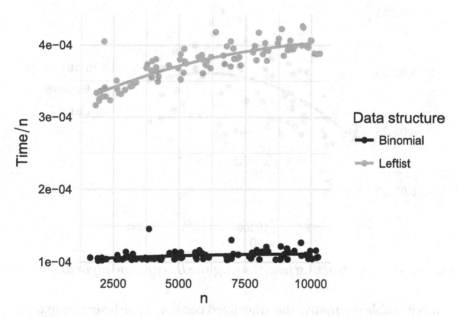

Figure 5-3. *Comparison of heap construction for leftist and binomial heaps when inserting one element at a time*

I should also stress that this benefit of using binomial heaps over leftist heaps is only relevant when we build heaps by inserting elements one at a time. If we use the algorithm that constructs heaps by iteratively merging larger and larger heaps, then both binomial and leftist heaps are constructed in linear time, and the leftist heap has a smaller overhead, as Figure 5-4, constructed using this experiment setup, shows:

```
setup <- function(n) n
evaluate <- function(empty) function(n, x) {
  vector_to_heap(1:n, empty)
}
```

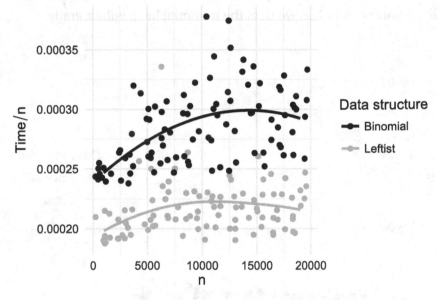

Figure 5-4. *Constructing heaps using the linear time algorithm*

It is possible to improve the amortized constant time insertion to a worst-case time by building the construction on skew binary numbers instead of binary numbers. This construction is similar to the random access list from chapter 3, and to the binomial heaps we have just implemented. All we need is to allow the two binomial trees with the smallest rank in a heap to have the same rank. If they do, we need to link them when we insert a new singleton tree, but just as we saw for the random access lists, by basing the construction on skew binary numbers rather than binary numbers, we never see a sequence of links being triggered by this. Of course, the difference between amortized and worst-case constant time operations is only relevant if every single operation needs to be fast—for example, for real-time or interactive applications—which is rarely a concern for algorithms in R, so we won't implement this variant here.

It is also possible to achieve worst-case merge operations by modifying the skew binomial heaps further into so-called Brodal heaps (Brodal and Okasaki 1996). This involves two levels of heaps. The Brodal heaps are skew binomial heaps containing *heaps* of the values we store in the heap, rather than the values themselves. Having two levels of heaps makes it possible to merge in constant time, but of course adds some complexity. However, with Brodal heaps all operations are constant time worst-case except delete minimal, which still has a logarithmic complexity. This is theoretically optimal in the sense that it is impossible to improve the complexity of the delete minimal operation without having to increase the complexity of some of the others.

Splay Heaps

A somewhat different approach to heaps is so-called *splay trees*. These are really search trees and have the search tree property—all values in a left subtree are smaller than the value in the root, and all values in the right subtree are larger—rather than the heap property. Because they are search trees, we will consider them in more detail in the next chapter, but here we will use them to implement a heap.

The structure for a splay tree is the same as for a search tree, and we saw this structure in Chapter 3. A search tree is defined recursively: it consists of a node with a value and a left and a right subtree that are also search trees. The invariant for search trees is the one mentioned earlier: the value in the node is larger than all values in the left subtree and smaller than all values in the right subtree. We can implement this structure like this:

```
splay_tree_node <- function(value, left = NULL, right = NULL) {
  list(left = left, value = value, right = right)
}
```

The structure we implemented in Chapter 3 used sentinel trees for empty trees. That was because we needed to do dispatch on generic methods on empty trees. For the splay heap we will implement now, we will not represent the heap as just a tree but wrap it in a structure that contains the tree and the minimal value in the heap, so, as with binomial heaps, we have a representation of empty heaps that don't rely on sentinels. Because of this, we will simply use NULL to represent empty trees.[6] The structure for splay heaps, the creation of empty heaps, and the test for emptiness are implemented like this:

```
splay_heap <- function(min_value, splay_tree) {
  structure(list(min_value = min_value, tree = splay_tree),
            class = c("splay_heap", "heap"))
}

empty_splay_heap <- function() splay_heap(NA, NULL)
is_empty.splay_heap <- function(x) is.null(x$tree)
```

---

[6]Using NULL to represent empty trees is a straightforward solution, but it does add some extra danger to working with the trees. In R, if you access a named value in a list that isn't actually in the list, you will get NULL back. This means, for example, that if you misspell a variable, say write x$lfet instead of x$left, you will not get any complaints when running the code, but you will always get NULL instead of what the real left subtree might be. Interpreting the default value when you have made an error as a meaningful representation of an empty tree is risky. We have to be extra careful when we do it.

We explicitly store the minimal value in the heap so we can return it in constant time:

```
find_minimal.splay_heap <- function(heap) {
  heap$min_value
}
```

To actually find the node with the minimal value in a search tree, we need to find the leftmost node in the tree. This is, by the search tree property, the smallest value. We can locate this node just by recursing to the left until we reach a node where the left child is empty:

```
splay_find_minimal_value <- function(tree) {
  if (is.null(tree)) NA
  else if (is.null(tree$left)) tree$value
  else splay_find_minimal_value(tree$left)
}
```

Here, we have a special case when the entire tree is empty—we use NA to represent the minimal value in that case. We should only hit this case if the entire heap is empty, though. Similar to finding the minimal value, we could find the maximum value by recursing to the right until the right subtree is empty.

This search takes time proportional to the depth of the tree. If we keep the tree balanced, then this would be $O(\log n)$, but for splay trees, we do not explicitly balance them. Instead, we blindly rearrange trees whenever we modify them.[7] Whenever we delete or insert values, we will do a kind of rebalancing that pulls the modified parts of the tree closer to the root.

---

[7]For a full splay tree implementation, we will also modify trees when we search in them. For the heap variant, where we explicitly store the minimal value, we do not need to do this, so for the splay heap, we only modify trees when we insert or remove from them.

We see some of this rearrangement in the code we use for deleting the smallest value in a tree. This value is the leftmost node in the tree, and a recursion that removes it has three cases: two basic cases that differ only in whether the leftmost node is the only left tree on the leftmost path or if it is the left tree of a left tree, and one recursive case, shown in Figure 5-5. In the first basis case, we can only return the right subtree. In the second, we also replace the leftmost node with its right subtree, so you can think of it as a special case. If we were just removing the leftmost tree, we wouldn't need this special case—we could just recurse to the left and use the result as the left subtree when returning from the recursion. The reason we need it is found in the recursive case. Here we need access to the root of a tree, its right subtree, its left subtree, and that subtree's two children. We will call recursively on the left subtree's left subtree and then rotate the tree, as shown in Figure 5-5. It is this rotation that has the special case when x is the minimal value in the tree that we handle as the second basic case.

Basic cases

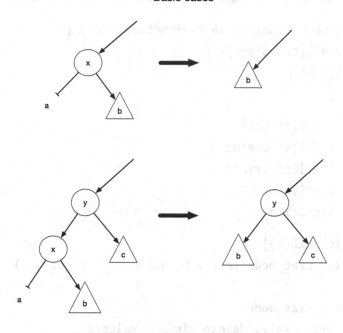

Recursive case

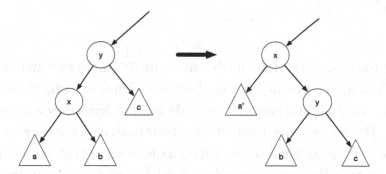

Figure 5-5. *Three cases in the recursion to delete the minimal node in a splay heap. In the recursive case, the tree a' refers to the result of calling recursively on tree a.*

Here is the code for deleting the minimal node in a splay tree:

```
splay_delete_minimal_value <- function(tree) {
  if (is.null(tree$left)) {
    tree$right

  } else {
    a <- tree$left$left
    x <- tree$left$value
    b <- tree$left$right
    y <- tree$value
    c <- tree$right

    if (is.null(a))
      splay_tree_node(left = b, value = y, right = c)
    else
      splay_tree_node(
        left = splay_delete_minimal_value(a),
        value = x,
        right = splay_tree_node(left = b, value = y, right = c)
      )
  }
}
```

The rotations we perform on the tree as we delete the minimal value do not balance the tree, but they do shorten the leftmost path. All nodes on the existing leftmost path will end up at half the depth they were before the call. The nodes in tree *b* will either get one node closer to the root or remain at their original depth, and nodes in tree *c* will either remain at their current depth or increase their depth by one node. We are not balancing the tree, but we are making the search path for the smallest value shorter on average, and it can be shown that we end up with an amortized $O(\log n)$ running time per delete_minimal operation if we use this approach.[8]

To delete the minimal value in the splay heap, we need to delete it from the underlying splay tree and then update the stored minimal value. We can implement this operation this way:

```
delete_minimal.splay_heap <- function(heap) {
  if (is_empty(heap))
    stop("Can't delete the minimal value in an empty heap")
  new_tree <- splay_delete_minimal_value(heap$tree)
  new_min_value <- splay_find_minimal_value(new_tree)
  splay_heap(min_value = new_min_value, splay_tree = new_tree)
}
```

When inserting a new element into a splay tree, we always put it at the root. To ensure the search tree property, we then have to put all elements smaller than the new value into the left subtree and all elements larger into the right subtree. To do that, we have a function, partition, that collects all the smaller and all the larger elements than a "pivot" element and returns them as splay trees. These "smaller" and "larger" trees are computed recursively based on the value in the current node and its left or right subtree. By looking at these two values, we can identify the subtree we need to recurse on to partition deeper parts of the tree, as shown in Figure 5-6. In the figure, $S(x)$ denotes the "smaller" tree we get by recursing on the tree x, and $L(x)$ denotes the "larger" tree we get by recursing on x.

---

[8]Proving this is mostly an exercise in arithmetic, and I will not repeat this analysis here. If you are interested, you can check Okasaki (1999a) or any text book describing splay trees.

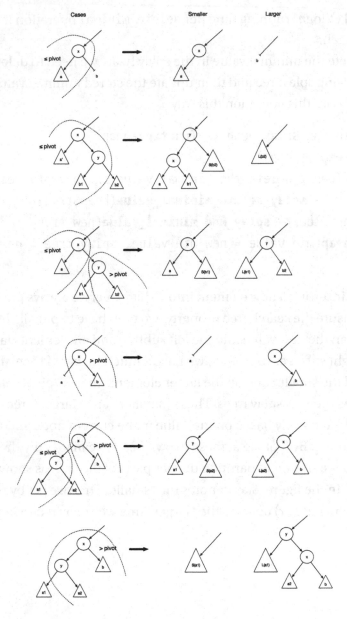

Figure 5-6. *Cases for splay heap partitioning (the special case where the tree is empty is left out)*

The implementation of partition is not particularly elegant, but it just considers the case of an empty tree first and then each of the six cases from Figure 5-6 in turn:

```
partition <- function(pivot, tree) {
  if (is.null(tree)) {
    smaller <- NULL
    larger <- NULL

  } else {
    a <- tree$left
    x <- tree$value
    b <- tree$right
    if (x <= pivot) {
      if (is.null(b)) {
        smaller <- tree
        larger <- NULL
      } else {
        b1 <- b$left
        y <- b$value
        b2 <- b$right
        if (y <= pivot) {
          part <- partition(pivot, b2)
          smaller <- splay_tree_node(
            left = splay_tree_node(
              left = a,
              value = x,
              right = b1
            ),
            value = y,
            right = part$smaller
          )
          larger <- part$larger
```

```r
    } else {
      part <- partition(pivot, b1)
      smaller <- splay_tree_node(
        left = a,
        value = x,
        right = part$smaller
      )
      larger <- splay_tree_node(
        left = part$larger,
        value = y,
        right = b2
      )
    }
  }
} else {
  if (is.null(a)) {
    smaller <- NULL
    larger <- tree
  } else {
    a1 <- a$left
    y <- a$value
    a2 <- a$right
    if (y <= pivot) {
      part <- partition(pivot, a2)
      smaller <- splay_tree_node(
        left = a1,
        value = y,
        right = part$smaller
      )
```

```
          larger <- splay_tree_node(
            left = part$larger,
            value = x,
            right = b
          )
        } else {
          part <- partition(pivot, a1)
          smaller <- part$smaller
          larger <- splay_tree_node(
            left = part$larger,
            value = y,
            right = splay_tree_node(
              left = a2,
              value = x,
              right = b
            )
          )
        }
      }
    }
  }
  list(smaller = smaller, larger = larger)
}
```

It is, unfortunately, not that unusual to have such long and inelegant functions for matching different cases in data structure manipulations. The cases themselves are not terribly complicated—we recognize a given shape of the tree and then we transform it into another shape—but implementing the tests and transformations can be very cumbersome and error prone.

We can improve the readability of the function somewhat by moving the cases' tests and transformations into separate functions. We can write a predicate function per case and a transformation we invoke whenever the predicate is satisfied:

```r
is_case_1 <- function(pivot, tree) {
  a <- tree$left
  x <- tree$value
  b <- tree$right
  x <= pivot && is.null(b)
}

transform_case_1 <- function(pivot, tree) {
  a <- tree$left
  x <- tree$value
  b <- tree$right
  list(smaller = tree, larger = NULL)
}

is_case_2 <- function(pivot, tree) {
  # is only called when right is not empty...
  a <- tree$left
  x <- tree$value
  b1 <- tree$right$left
  y <- tree$right$value
  b2 <- tree$right$right
  x <= pivot && y <= pivot
}

transform_case_2 <- function(pivot, tree) {
  # is only called when right is not empty...
  a <- tree$left
  x <- tree$value
  b1 <- tree$right$left
```

```
  y <- tree$right$value
  b2 <- tree$right$right

  part <- partition(pivot, b2)
  smaller <- splay_tree_node(
    left = splay_tree_node(
      left = a,
      value = x,
      right = b1
    ),
    value = y,
    right = part$smaller
  )
  larger <- part$larger

  list(smaller = smaller, larger = larger)
}

is_case_3 <- function(pivot, tree) {
  # is only called when right is not empty...
  a <- tree$left
  x <- tree$value
  b1 <- tree$right$left
  y <- tree$right$value
  b2 <- tree$right$right
  x <= pivot && y > pivot
}

transform_case_3 <- function(pivot, tree) {
  # is only called when right is not empty...
  a <- tree$left
  x <- tree$value
  b1 <- tree$right$left
```

```
  y <- tree$right$value
  b2 <- tree$right$right

  part <- partition(pivot, b1)
  smaller <- splay_tree_node(
    left = a,
    value = x,
    right = part$smaller
  )
  larger <- splay_tree_node(
    left = part$larger,
    value = y,
    right = b2
  )

  list(smaller = smaller, larger = larger)
}

is_case_4 <- function(pivot, tree) {
  a <- tree$left
  x <- tree$value
  b <- tree$right
  x > pivot && is.null(a)
}

transform_case_4 <- function(pivot, tree) {
  a <- tree$left
  x <- tree$value
  b <- tree$right
  list(smaller = NULL, larger = tree)
}
```

```r
is_case_5 <- function(pivot, tree) {
  # is only called when left is not empty
  a1 <- tree$left$left
  y <- tree$left$value
  a2 <- tree$left$right
  x <- tree$value
  b <- tree$right
  x > pivot && y <= pivot
}

transform_case_5 <- function(pivot, tree) {
  # is only called when left is not empty
  a1 <- tree$left$left
  y <- tree$left$value
  a2 <- tree$left$right
  x <- tree$value
  b <- tree$right

  part <- partition(pivot, a2)
  smaller <- splay_tree_node(
    left = a1,
    value = y,
    right = part$smaller
  )
  larger <- splay_tree_node(
    left = part$larger,
    value = x,
    right = b
  )

  list(smaller = smaller, larger = larger)
}
```

```r
is_case_6 <- function(pivot, tree) {
  # is only called when left is not empty
  a1 <- tree$left$left
  y <- tree$left$value
  a2 <- tree$left$right
  x <- tree$value
  b <- tree$right
  x > pivot && y > pivot
}

transform_case_6 <- function(pivot, tree) {
  # is only called when left is not empty
  a1 <- tree$left$left
  y <- tree$left$value
  a2 <- tree$left$right
  x <- tree$value
  b <- tree$right

  part <- partition(pivot, a1)
  smaller <- part$smaller
  larger <- splay_tree_node(
    left = part$larger,
    value = y,
    right = splay_tree_node(
      left = a2,
      value = x,
      right = b
    )
  )

  list(smaller = smaller, larger = larger)
}
```

This approach is slightly less efficient because we extract parts of the tree repeatedly in both tests and transformations and we repeat tests in different predicates, but the code is clearer and more manageable. Clear code is preferable to a complex sequence of if-statements, so unless performance *really* becomes a problem, the cleaner version is the better one.[9] With these predicates and transformations, the `partition` function just tests and transforms:

```
partition <- function(pivot, tree) {
  if (is.null(tree))
    list(smaller = NULL, larger = NULL)
  else if (is_case_1(pivot, tree))
    transform_case_1(pivot, tree)
  else if (is_case_2(pivot, tree))
    transform_case_2(pivot, tree)
  else if (is_case_3(pivot, tree))
    transform_case_3(pivot, tree)
  else if (is_case_4(pivot, tree))
    transform_case_4(pivot, tree)
  else if (is_case_5(pivot, tree))
    transform_case_5(pivot, tree)
  else if (is_case_6(pivot, tree))
    transform_case_6(pivot, tree)
  else stop("Unknown case")
}
```

With `partition` in place, implementing `insert` is straightforward: we partition on the element we insert, put the smaller values to the left of the heap, the larger elements to the right, and the new element at the root, and

---

[9]If performance becomes enough of a problem that the long version of `partition` is needed, that performance is probably better achieved by moving some of the code to C/C++ than by having the complicated `partition` function.

then we updated the minimal value if the new element is smaller than the previous minimal value:

```
insert.splay_heap <- function(x, elm) {
  part <- partition(elm, x$tree)
  new_tree <- splay_tree_node(
    value = elm,
    left = part$smaller,
    right = part$larger
  )
  new_min_value <- min(x$min_value, elm, na.rm = TRUE)
  splay_heap(min_value = new_min_value, splay_tree = new_tree)
}
```

Merging splay heaps is less efficient than merging the other heaps we have seen. To merge two splay heaps, we partition on the root of one of the trees, put that root at the root of the new tree, and then merge the left and right subtrees into the smaller and larger parts of the partition, recursively. The whole operation runs in time $O(n)$ worst case:

```
merge_splay_trees <- function(x, y) {
  if (is.null(x)) return(y)
  if (is.null(y)) return(x)

  a <- x$left
  val <- x$value
  b <- x$right

  part <- partition(val, y)
  splay_tree_node(left = merge_splay_trees(part$smaller, a),
                  value = val,
                  right = merge_splay_trees(part$larger, b))
}
```

```
merge.splay_heap <- function(x, y, ...) {
  if (is_empty(x)) return(y)
  if (is_empty(y)) return(x)

  new_tree <- merge_splay_trees(x$tree, y$tree)
  new_min_value <- min(x$min_value, y$min_value, na.rm = TRUE)
  splay_heap(min_value = new_min_value, splay_tree = new_tree)
}
```

Best-case performance for both insertion and merging is cases where partition completes early and happens if we insert elements in a splay heap in order or when we merge two heaps where all the elements in one are larger than all the elements in the other. If we build a heap from sorted elements, or elements sorted in reverse order, inserting one element at a time, splay heaps outperforms binomial heaps, as shown in Figure 5-7. And if we build heaps by iteratively merging smaller heaps and start with the elements ordered, splay heaps are competitive to leftist heaps, as shown in Figure 5-8.

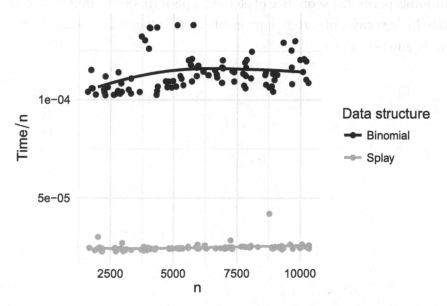

Figure 5-7. *Constructing splay heaps one element at a time*

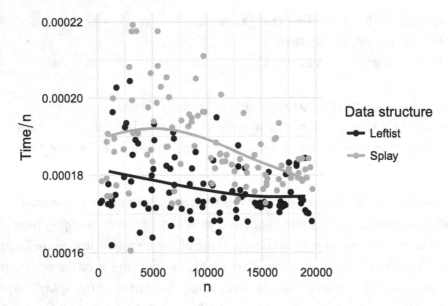

Figure 5-8. *Constructing splay heaps by iteratively merging heaps*

In both construction algorithms, though, constructing heaps from a randomly permuted sequence of elements performs substantially worse than the best cases of sorted or reversely sorted elements, as Figure 5-9 and Figure 5-10 show.

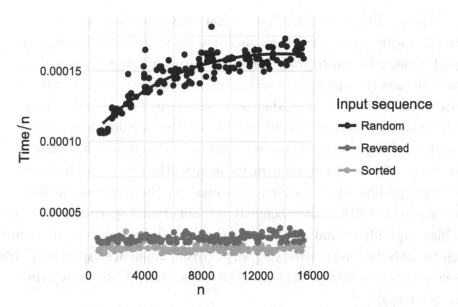

Figure 5-9. *Constructing splay heaps one element at a time for different types of input*

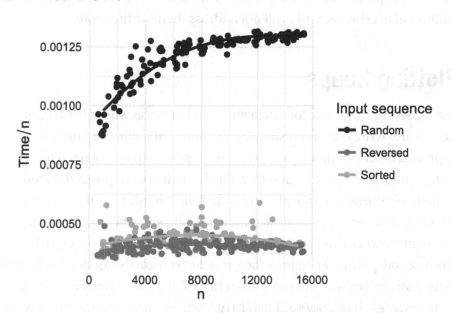

Figure 5-10. *Constructing splay heaps by iteratively merging for different types of input*

These experiments make it seem that splay heaps are ideal for working with sequences that are almost in order, and when it comes to performance for *constructing* heaps, this is true. It comes at a cost, though. We can construct splay heaps very quickly because we do not balance them. This will cost us later when we have to search down in the heap. The splay operation will rebalance the heap when we do that, so we will not have a large overhead because of the lack of balance—after a few deletions of the minimal element, the heap will be balanced—but it *can* give you problems with the recursive functions. There is a limit to how deep you can call functions before R runs out of stack space and gives up. This is typically around 1,000 function instantiations, so about a thousand elements is the limit of what you can put in an unbalanced splay heap. The two other heaps that are kept balanced will not have problems with much larger heaps.

There are ways around this recursion problem (see Mailund 2017a for a general approach based on so-called *trampolines*), but these come with additional overheads, and I will not address them further here.

Plotting Heaps

Just as we can write code for displaying search trees, as we did in Chapter 3, we can also write code for displaying the different heap structures. The approach is very similar to what we saw for search trees: we need to collect the graph structure and then display it. It is even possible to reuse some code; the search tree plotting code from Chapter 3 generalizes to all binary trees and can be reused for leftist and splay heaps, and with a little more work can be generalized to binomial heaps. Most of such plotting code, however, follows the same pattern and is not directly related to data structures, so I will not show it here. For those interested, I refer to `https://github.com/mailund/ralgo`, where the data structures are implemented together with display code. In the next chapter, I will show

how to extend the plotting code for search trees to annotate a tree with additional information. For the heaps, I will simply show the result of some operations, and if you want to experiment yourself, you will have to download the `ralgo` package:

```
if(!require(devtools)) install.packages("devtools")
devtools::install_github("mailund/ralgo")
```

For the three heaps we have implemented, I will construct one of size 10 and then delete three elements, plotting the results along the way:

```
heap <- empty_leftist_heap()
for (i in 1:10)
  heap <- insert(heap, i)
plot(heap)

heap <- delete_minimal(heap)
plot(heap)

heap <- delete_minimal(heap)
plot(heap)

heap <- delete_minimal(heap)
plot(heap)
```

The results for a leftist heap are shown in Figure 5-11. The structure is relatively straightforward, and we see that this heap is balanced in all the four plots and that the trees satisfy the heap property.

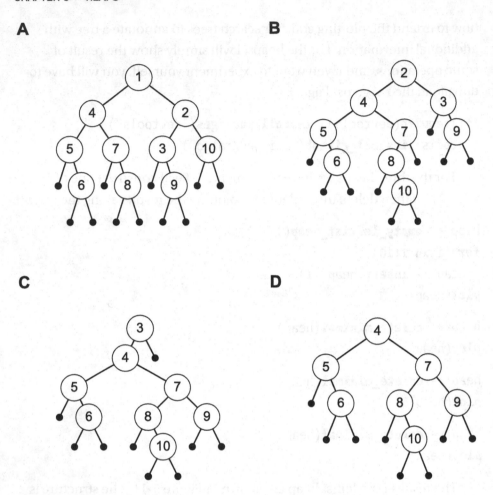

Figure 5-11. *Leftist heaps*

Binomial heaps are shown in Figure 5-12. Here the black nodes display the list of trees in the heap, and the numbers in the black nodes are the rank of the trees in these links, not values from the heap.

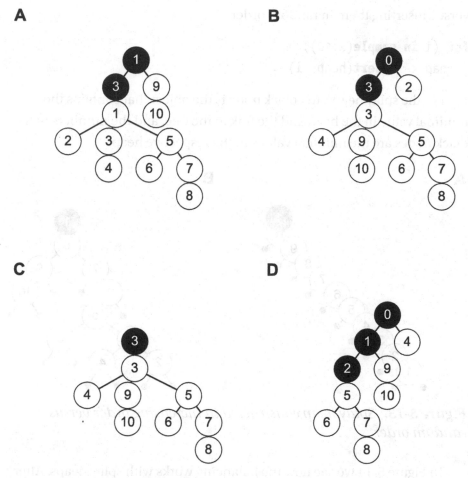

Figure 5-12. *Binomial heaps*

Splay heaps are not explicitly kept balanced, so the initial heap will depend on the order in which we insert elements. Figure 5-13 shows the difference between inserting the elements in increasing order

```
for (i in 1:10)
  heap <- insert(heap, i)
```

versus inserting them in random order

```
for (i in sample(1:10))
  heap <- insert(heap, i)
```

For the splay heaps, the black node is the object that contains the minimal value in the heap and the link to the tree, and the numbers in the black nodes are the minimal values in the respective heaps.

A   **B**

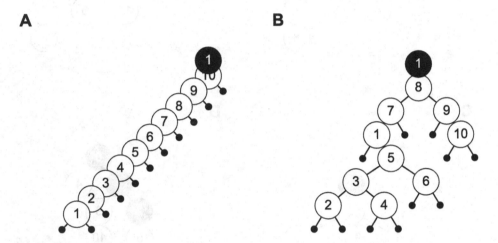

Figure 5-13. *Splay heap constructed in increasing order versus random order*

In Figure 5-14 we see how the balancing works with splay heaps. After inserting elements in increasing order, the heap is maximally unbalanced, but as we delete elements, the splay operation transforms the tree into a more balanced structure.

A

B

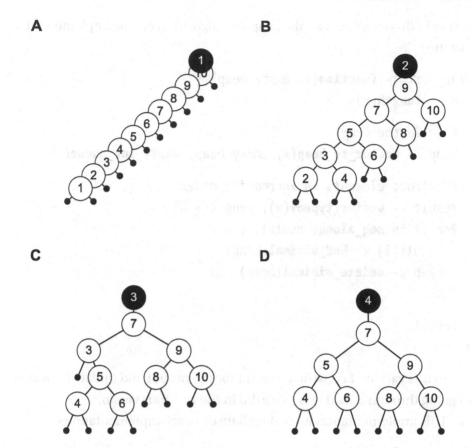

C

D

Figure 5-14. *Splay heaps*

Heaps and Sorting

To see the heaps in action, we can revisit the heap sort algorithm discussed earlier in this chapter. The idea is simply this: if we want to sort a sequence of elements, we can build a heap from them and then iteratively extract the minimal element. If we use the vector_to_heap function we implemented earlier in this chapter, then the heap sort algorithm can be implemented like this:

```
heap_sort <- function(x, empty_heap) {
  n <- length(x)

  # Build the heap
  heap <- vector_to_heap(x, empty_heap, empty_env_queue())

  # extract elements in increasing order
  result <- vector(typeof(x), length = n)
  for (i in seq_along(result)) {
    result[i] <- find_minimal(heap)
    heap <- delete_minimal(heap)
  }

  result
}
```

In this variant, I construct a vector for the output and update it instead of going through linked lists as we did in the previous version.

This implementation works for all three heap implementations:

```
(x <- sample(10))
## [1]  3  7  8  2  1  4 10  5  9  6
heap_sort(x, empty_leftist_heap())
## [1]  1  2  3  4  5  6  7  8  9 10
heap_sort(x, empty_binomial_heap())
## [1]  1  2  3  4  5  6  7  8  9 10
heap_sort(x, empty_splay_heap())
## [1]  1  2  3  4  5  6  7  8  9 10
```

The running time differs between them, however. For both the leftist heap and the binomial heap, we can construct the heap in linear time and then spend $O(n\log n)$ time extracting the elements one at a time.

Iteratively merging splay heaps in vector_to_heap is worst-case an $O(n^2)$ algorithm, though, where iteratively inserting an element would be better at expected $O(n\log n)$. It turns out that it is inherently more complex to construct splay heaps than the two other heaps, and it cannot be done faster than $O(n\log n)$. To see this, you must know that it is impossible to sort n elements, based on comparisons to determine the order, faster than $O(n\log n)$. Further, if you have a search tree, you are able to extract elements in order via a depth first traversal in linear time. We can modify the sort algorithm for splay trees to extract the elements in this way, so constructing result is done in $O(n)$. This immediately tells us that constructing the heap must take at least $O(n\log n)$ because otherwise we would be sorting faster than this.

A sort function tailored to the splay heap could look like this:

```
splay_heap_sort <- function(x) {
  n <- length(x)

  heap <- empty_splay_heap()
  for (element in x)
    heap <- insert(heap, element)

  result <- vector(typeof(x), length = n)
  i <- 1
  recurse <- function(tree) {
    if (!is_empty(tree)) {
      recurse(tree$left)
      result[i] <<- tree$value
      i <<- i + 1
      recurse(tree$right)
    }
  }
```

```
  recurse(heap$tree)
  result
}
```

```
splay_heap_sort(x)
## [1]  1  2  3  4  5  6  7  8  9 10
```

The imperative access to result and the use of the variable i are not pretty, but the immutability of function arguments makes it hard to work with vectors in recursive tree traversals without something similar to this. We could avoid it by building a list in the traversal instead—and traverse the tree with recursions to the right instead of left to get the list constructed in the right order—but that would just move the problem to the function we need to translate the list into a vector where we would still need an index variable that has to be updated. In any case, we have encapsulated the ugliness in the function, so at least users of splay_heap_sort cannot see the side effects we have in the recursion.

In Figure 5-15 we see comparisons of the three variants of heap sorting on data that is already sorted or that is in a random order before we call the sort function. The leftist and binomial heaps have practically the same running time, but the splay heap is very sensitive to the order of the input sequence. If the sequence is already ordered, it can construct the heap very quickly and then exploit that it only takes linear time to traverse the tree; if the input is in random order, it suffers from being unbalanced in the heap construction. For the ordered input sequence, the recursive calls in the splay heap prevent us from inserting more than about a thousand elements, though, so although the performance is superior for this combination of splay heaps and sequence order, it does have its problems unless you take care to avoid the function stack problems.

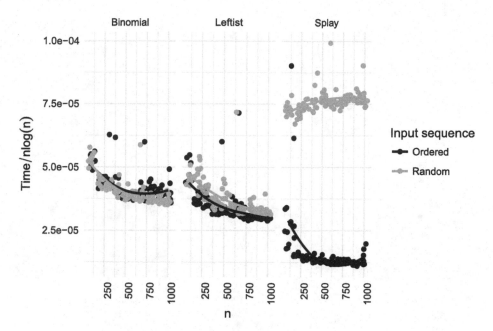

Figure 5-15. *Heap sort running time*

CHAPTER 6

Sets and Search Trees

We saw earlier how we can represent collections of elements such that we can efficiently iterate through them and efficiently merge two such collections. In this chapter, we turn "bags" into "sets" by considering data structures that allow us to efficiently insert, delete, and check for membership of elements:

```
insert <- function(x, elm) UseMethod("insert")
remove <- function(x, elm) UseMethod("remove")
member <- function(x, elm) UseMethod("member")
```

We do this through so-called *search trees*—trees with the property that all elements in the left subtree of a node will have values smaller than the element in the node and all elements in the right subtree will have values larger than the value in the node. This property makes it easy to search for any given element in a tree and thus test for membership. It also gives us a way of implementing mappings from keys to values: we just need to have a search tree for keys and also store the corresponding values in the nodes together with the keys.

© Thomas Mailund 2017
T. Mailund, *Functional Data Structures in R*, https://doi.org/10.1007/978-1-4842-3144-9_6

Search Trees

In Chapter 3, we implemented a basic, unbalanced search tree like this:

```
search_tree_node <- function(
  value
  , left = empty_search_tree()
  , right = empty_search_tree()
) {
  structure(list(left = left, value = value, right = right),
            class = c("unbalanced_search_tree"))
}

empty_search_tree <- function()
  search_tree_node(NA, NULL, NULL)
is_empty.unbalanced_search_tree <- function(x)
  is.null(x$left) && is.null(x$right)
```

with this straightforward membership test

```
member.unbalanced_search_tree <- function(x, elm) {
  if (is_empty(x)) return(FALSE)
  if (x$value == elm) return(TRUE)
  if (elm < x$value) member(x$left, elm)
  else member(x$right, elm)
}
```

or the slightly more complex

```
st_member <- function(x, elm, candidate = NA) {
  if (is_empty(x)) return(!is.na(candidate) && elm ==
candidate)
  if (elm < x$value) st_member(x$left, elm, candidate)
  else st_member(x$right, elm, x$value)
}
```

```
member.unbalanced_search_tree <- function(x, elm) {
    st_member(x, elm)
}
```

with a simple insertion function that just performs the search and creates an updated tree along the way:

```
insert.unbalanced_search_tree <- function(x, elm) {
  if (is_empty(x)) return(search_tree_node(elm))
  if (elm < x$value)
    search_tree_node(x$value, insert(x$left, elm), x$right)
  else if (elm > x$value)
    search_tree_node(x$value, x$left, insert(x$right, elm))
  else
    x # the value is already in the tree
}
```

The removal function was slightly more complicated. Deleting elements that are leaves or have only a single child is simple, but most elements are not stored like that. So we used a trick where we replaced the element we had to remove with the leftmost element in the right subtree—we can replace values much easier than we can delete values—and then removed the leftmost tree in the right subtree. Because leftmost values are either roots or have only a right subtree, this reduces the removal to one of the simple cases we can deal with directly:

```
st_leftmost <- function(x) {
  while (!is_empty(x)) {
    value <- x$value
    tree <- x$left
  }
  value
}
```

```r
remove.unbalanced_search_tree <- function(x, elm) {
  # if we reach an empty tree, there is nothing to do
  if (is_empty(x)) return(x)

  if (x$value == elm) {
    a <- x$left
    b <- x$right
    if (is_empty(a)) return(b)
    if (is_empty(b)) return(a)

    s <- st_leftmost(x$right)
    return(search_tree_node(s, a, remove(b, s)))
  }

  # we need to search further down to remove the element
  if (elm < x$value)
    search_tree_node(x$value, remove(x$left, elm), x$right)
  else # (elm > x$value)
    search_tree_node(x$value, x$left, remove(x$right, elm))
}
```

Operations on search trees take time proportional to the depth we need to reach in the operations, so they are bounded by the depth of the tree, which if we do nothing to balance trees can be $O(n)$. For balanced trees, though, the runtime complexity of the operations are only $O(n\log n)$, so it pays to keep the trees balanced, and that is the trick we need to address in this chapter.

Red-Black Search Trees

Red-black search trees are kept balanced by imagining that we color all nodes either red or black and keeping the following invariants:

1. No red node has a red parent.

2. Every path from the root to a leaf has the same
    number of black nodes.

The second invariant guarantees that the tree is balanced if we only consider the black nodes. Any path from the root to a leaf will go through exactly the same number of black nodes, so measured in the depth of black nodes, the tree cannot be deeper than $O(\log n)$. Not all paths from the root to a leaf necessarily have the exact same length because there are also the red nodes, but the first invariant guarantees that we can't have more red nodes along a path than we have black nodes. Consequently, the longest path from the root to a leaf is no more than twice the length of the shortest path, so we are still guaranteeing a logarithmic depth of all leaves.

To guarantee the invariants, we need to modify the insertion and deletion operations. Searching can remain the way it is. For insertion, this turns out to be a simple change (Okasaki 1999b) whereas deletion is somewhat more complicated (Germane and Might 2014).

We implement the data structure by storing the color for each node in the node, and we use a sentinel for empty trees:

```
# colours
RED <- 1
BLACK <- 2

# helper function
red_black_tree_node <- function(
  colour
  , value
  , left = empty_red_black_tree()
  , right = empty_red_black_tree()
  ) {
  structure(list(colour = colour,
                 left = left,
```

```
                 value = value,
                 right = right),
             class = "red_black_tree")
}
empty_red_black_tree <- function()
    red_black_tree_node(BLACK, NA, NULL, NULL)

is_empty.red_black_tree <- function(x)
    is.null(x$left) && is.null(x$right)
```

We put empty trees as the children of leaves or as one of the children for a node with only one child, and we arbitrarily assign the color black to those. All paths leading down to an empty tree will end with a black node, so this doesn't affect the second invariant—it just makes all paths one black node longer—and it avoids having to worry about inserting red leaves that would then be parents of a red empty tree.

The structure for a red-black search tree is the same as the unbalanced tree, except for the color information. Thus, searching in the tree works exactly the same. Because both can use the same method, we can implement it for the superclass search_tree and simply let them both inherit it:

```
member.search_tree <- function(x, elm) {
  st_member(x, elm)
}
```

This method uses the slightly faster membership function we saw in Chapter 3.

We just need to give both the classes this superclass, so we modify their constructors as follows:

```
search_tree_node <- function(
  value
  , left = empty_search_tree()
  , right = empty_search_tree()
```

```
) {
  structure(list(left = left, value = value, right = right),
           class = c("unbalanced_search_tree", "search_tree"))
}

red_black_tree_node <- function(
  colour
  , value
  , left = empty_red_black_tree()
  , right = empty_red_black_tree()
  ) {
  structure(list(colour = colour,
                 left = left,
                 value = value,
                 right = right),
           class = c("red_black_tree", "search_tree"))
}
```

Insertion

Now, the trick to inserting elements and keeping the tree balanced is to rearrange the tree after an insertion. We insert new elements exactly as we did for the unbalanced tree: we search down in the tree until we find the position where the new node should be inserted, and then we insert a new, red leaf for the element. In the recursion, going down the tree in the search, we call a function that is responsible for rebalancing the tree; this function looks at a local part of the tree and rearranges it to keep the invariants true. The insertion looks like this:

```
rbt_insert <- function(tree, elm) {
  if (is_empty(tree)) return(red_black_tree_node(RED, elm))
  if (elm < tree$value)
```

```
    rbt_balance(tree$colour,
                tree$value,
                rbt_insert(tree$left, elm),
                tree$right)
  else if (elm > tree$value)
    rbt_balance(tree$colour,
                tree$value,
                tree$left,
                rbt_insert(tree$right, elm))
  else
    tree # the value is already in the tree, at this level,
         # so just return
}

insert.red_black_tree <- function(x, elm, ...) {
  # insert the value in the tree and set the root to be black
  new_tree <- rbt_insert(x, elm)
  new_tree$colour <- BLACK
  new_tree
}
```

In the insert function we call rbt_insert and color the root of the tree black. Changing the root color to black will automatically satisfy both invariants if the children of the root satisfy them since it potentially adds a black node to *all* paths, satisfying invariant one, and cannot invalidate invariant two even if the root's children are red.

The rbt_insert function will insert a red leaf when it hits the end of the search but otherwise just searches recursively, making sure to call a function for balancing, rbt_balance, for each node it traverses going down the tree in the search.

As for the balancing function, we handle local parts of the tree by considering various cases the structure of the tree can be in after we have updated it. When we insert a new leaf, we insert it as red. This cannot invalidate the second invariant, but it can invalidate the first. As we return from the recursion, we will consider the cases where we have a red parent to a red child and rearrange the tree to avoid this, while still guaranteeing that we do not change the number of black nodes on any path.

The transformations we need to perform in the balancing are shown in Figure 6-1. There are four possible cases where a red node has a red parent, shown in the four trees at the edges of the figure, and in all cases, we transform the tree into the structure shown at the middle of the figure. If the two invariants are valid in all other parts of the tree, they will also be valid after the transformation, and we will have removed the red node with a red parent. To see this, we can observe that the red node with the red parent is removed in the middle tree, and if we consider any path going through this part of the tree, there is one black node at the root of all the trees before the transformation. Since we replace the black root with a red one and make its two children black, all paths going through this part of the tree will also see a single black node. After the transformation, we *might* have placed a new red node as the child of another red node—we have made the root of this subtree red—but this will then be taken care of as we continue up the recursion.

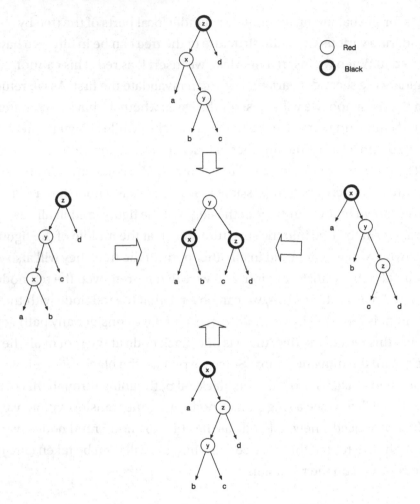

Figure 6-1. *Transformations for balancing a red-black search tree*

The `rbt_balance` function just checks the four cases and then performs the relevant transformation. When we implemented splay heaps, we did the case checking using two functions per case, one for checking the case and one for the transformation. For the red-black search tree, the transformations are a little simpler: all the cases transform to the same case, so if we simply name the nodes and subtrees with variables as shown in the figure, then we can reuse the transformation—but of course then we need the transformation to have access to the variables as defined by the case tests. For the case analysis in our implementation of `rbt_balance`, we are going to use a different strategy than for the splay heaps. The implementation of this trick is a lot more technical than simply having functions for checking the cases, but the use of the trick is correspondently simpler, so the trick is worth considering when implementing transformations. And you can just reuse the implementation from this chapter for future projects—it should generalize to other case analyses as long as you follow a few rules.

We are going to check for the structure of the local tree, and bind nodes and subtrees to variables at the same time, using a bit of non-standard evaluation. If you are not familiar with this, you can read more about it in my book *Metaprogramming in R* (Apress, 2017) or Hadley Wickham's *Advanced R* (Chapman & Hall, 2014). The trick is to evaluate parameters we provide to a function as if they are either assignments, which we do when we have named parameters, or as logical expressions when they are just positional arguments.

The function, `pattern_match`, is shown in the next code block. It first takes all its arguments, using the expression `eval(substitute(alist(...)))`, which is the standard approach to getting a list of unvalued parameters, and then gets the calling scope using `parent.frame()`. We need to evaluate the parameters in the calling scope, rather than the local scope of `pattern_match`, so we can refer to variables in the scope where we use the function.

Next, we get the names of the parameter bindings. These will be non-empty strings when we provide named arguments and empty strings otherwise. We now iterate through all the arguments of the function and evaluate expressions provided to the arguments in the calling scope. If any value evaluates to `NULL`, we terminate the function and return `FALSE`. If the argument was named, we assign the value to the name in the calling scope, but if the argument is not named, we check whether it is evaluated to a true value. If not, we again terminate and return FALSE. If we make it through all the arguments without seeing values NULL or FALSE, we return TRUE:

```
pattern_match <- function(...) {
  bindings <- eval(substitute(alist(...)))
  scope <- parent.frame()

  var_names <- names(bindings)
  for (i in seq_along(bindings)) {
    name <- var_names[i]
    val <- eval(bindings[[i]], scope)

    if (is.null(val)) return(FALSE)

    # for expressions that are not assignments,we consider them
    # conditions that must be true for the pattern to match.
    # Return FALSE if they are not.
    if (nchar(name) == 0 && (is.na(val) || !val)) return(FALSE)
    else if (nchar(name) > 0) assign(name, val, envir = scope)

  }
  return(TRUE)
}
```

If you do not understand all the details of this function, do not worry. Using the function is a lot easier than understanding the details of what it does.

The way we will use it is that we will check the structure of the tree by evaluating expressions that pick out subtrees and assign these subtrees to variables at the same time, and we will check the colors of nodes using logical expressions. For example, to check that we have the topmost case from Figure 6-1, we will use the expression shown in the next code block, in a context where the color of the root node is stored in the variable colour, the value in the top node is referred to by value, and the left and right subtrees are stored in variables left and right, respectively. The variables a, b, and c are assigned the subtrees shown in the figure, and the variables x, y, and z will be assigned the values in the nodes shown in the figure. The last three arguments to the function call are logical expressions, and these are just used to check that the colors of the nodes match those in the figure:

```
pattern_match(a = left$left, b = left$right$left,
              c = left$right$right, d = right,
              x = left$value, y = left$right$value, z = value,
              colour == BLACK, left$colour == RED,
              left$right$colour == RED)
```

Important for this function to perform correctly is that we do not use NULL to refer to empty trees, as we did for the splay heap. Whenever we use an expression such as left$left, if the list left doesn't have a named index left, then the expression would give us NULL. If we used NULL to represent empty trees, we would not be able to distinguish between an empty tree and a tree that isn't represented because we refer to it through a list that doesn't contain the named tree. In our implementation of red-black search trees, we represent empty trees as sentinel objects, so we *can* distinguish between empty trees and non-existing trees, and we exploit that by considering NULL as an indication that the tree we are looking at does not have the right structure.

We can see the function in action in the rbt_balance function. It checks whether any of the four patterns we need to transform are matched by the current tree. If so, we transform according to the rules in Figure 6-1. If not, we just leave the tree as it is:

```
rbt_balance <- function(colour, value, left, right) {
  # Setting these to avoid warnings
  a <- b <- c <- d <- x <- y <- z <- NULL
  if (pattern_match(a = left$left, b = left$right$left,
```

```
                        c = left$right$right, d = right,
                        x = left$value, y = left$right$value,
                        z = value,
                        colour == BLACK, left$colour == RED,
                        left$right$colour == RED)

    || pattern_match(a = left$left$left, b = left$left$right,
                        c = left$right, d = right,
                        x = left$left$value, y = left$value,
                        z = value,
                        colour == BLACK, left$colour == RED,
                        left$left$colour == RED)

    || pattern_match(a = left, b = right$left,
                        c = right$right$left,
                        d = right$right$right,
                        x = value, y = right$value,
                        z = right$right$value,
                        colour == BLACK, right$colour == RED,
                        right$right$colour == RED)

    || pattern_match(a = left, b = right$left$left,
                        c = right$left$right, d = right$right,
                        x = value, y = right$left$value,
                        z = right$value,
                        colour == BLACK, right$colour == RED,
                        right$left$colour == RED)
) {

  left <- red_black_tree_node(colour = BLACK, value = x,
                              left = a, right = b)
  right <- red_black_tree_node(colour = BLACK, value = z,
                               left = c, right = d)
```

```
    red_black_tree_node(colour = RED, value = y, left, right)
  } else {
    red_black_tree_node(colour, value, left, right)
  }
}
```

Whether you prefer the approach we took with splay heaps, where we did the case analysis and transformations using functions, or the pattern matching we used here is a matter of taste. Because the actual transformation is the same for all four cases here, except that the variables used in the transformation refer to different trees and values in the tree before the transformation, I prefer the pattern matching approach for this function.

Deletion

When it comes to deleting elements in a red-black search tree, the number of transformations necessary to preserve the invariants grows a bit. The basic idea behind deletion is the same as for the unbalanced search tree: it is easy to remove elements in leaves or in nodes with only a single child, so those we can delete directly. For inner nodes, we will switch the value with its successor in the tree—the leftmost value in the right subtree—and then delete that value from the right subtree. If the node we delete is red, we will not invalidate any of the invariants. If it is black, on the other hand, we are removing one black node from a path to a root, which will invalidate the second invariant. To deal with this, we need transformations of the tree similar to the rbt_balance transformations.

When we delete a black node with zero or one children, we will have removed a black node from a path from the root down to one or more leaves—either in the subtree that the node we delete had or in its parent if it is now a leaf. This invalidates the first invariant, but we cannot do much about it locally in the tree where we delete the node. Instead, we will use an additional color—we call it *double black*—that works as if we had two black nodes when we consider a path going through it. We will use this color when we delete a black node, and then we will let it double up towards the root until we can find a local tree configuration where we can put its children one black node higher in the tree than other paths, thus evening out the balance. We will see how we do this shortly:

```
DOUBLE_BLACK <- 3
```

First, let us just handle the cases where we can delete a node directly. These are shown in Figure 6-2, where v refers to the value we want to delete. I have shown the empty trees we use as sentinels explicitly here. For the double black color to do its trick, we need to consider these explicitly. There are four cases: we can have v in a red leaf, in a black node with one red child—there are two, symmetric cases—or we can have v in a black leaf. The first three we can handle without any problems; in the last case we are removing a black node, so we leave behind a double black empty tree.

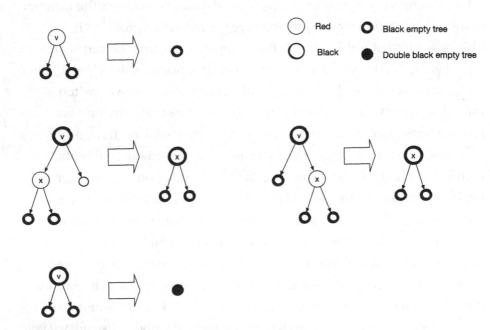

Figure 6-2. *Trivial deletions in a red-black search trees*

We now need to get rid of the double black node. We do this by considering various cases where this node can appear in the tree and either get rid of it or move it up the tree—if it eventually reaches the root, we can always color it black and get rid of it that way. We call these operations *rotations* to distinguish them from the balancing function we used when inserting elements.

Because we have already handled cases where at least one child of a node is empty, all the cases we need to consider are nodes with two children. The first invariant guarantees that we do not have two red nodes in a row, and we only consider cases where the double black node is one of the children, so all the cases are shown in Figure 6-3—although, here, the tree with the double black root is allowed to be empty and thus its children to actually be NULL. The non-double black tree cannot be empty. When we translate a black leaf into a double black empty tree, that black leaf must have had a black sibling—otherwise, the second invariant would have been violated. As long as we move the double black node up the tree, the invariant will be violated, and there must be a black node more on all paths in the double black nodes sibling-tree, thus this tree cannot be empty.

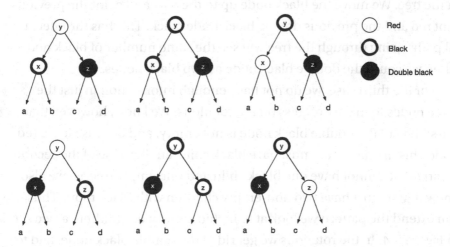

Figure 6-3. *Cases for red-black search tree rotations*

The actual rotations are shown in Figure 6-4. Notice here that the tree with the double black node is never rearranged. The color of the double black node might be changed, but the tree always keeps its two children. This means that we, when pattern matching, do not need to access its children, so when the double black node is the root of an empty tree—and these children would be NULL—we will not have problems with `pattern_match` considering this is not a match to the pattern. We modify the structure of the double black node's sibling, but as we have just argued, this sibling is never an empty tree.

There are three types of rearrangement, which makes six rules because of symmetry. In the first transformation, the root of the tree we consider is red, and the sibling of the double black node is (necessarily) black. In this case, we can move the black node up to the root of the tree and color the double black node black. By making the black node a grandparent of the double black node, we add a black node to the paths going down to the leaves in this subtree, compensating for the missing black node we represented by the double black node. In these rotations, we get rid of the double black node.

In the second case, both the root and the sibling are black. We cannot get rid of the double black node in this case, but we can push it upwards in the tree. We move the black node up to the root and color the previous root red and the previous double black node black. This has the effect that all paths going through the tree will see the same number of black nodes when we count the double black node as two black nodes.

For the third case, we do not have enough information in just the three nodes to make progress on the rotations. We know, however, that the sibling of the double black node is not empty, and because it is a red node, this means that it must have black children. Because of the second invariant, it cannot have one black child and one empty tree, so we also know that it must have two non-empty children with black roots. Thus, we can extend the pattern we look at to include one of its children, as we see in Figure 6-4. In the rotations we get rid of the double black node, and to see that the rotation is correct we just need to convince ourselves that the

subtrees and nodes are in the same in-order as before and that, if we count
the number of black nodes as we go through any path in the first tree to
one of the *a* to *e* trees, counting the double black node as two black nodes,
we will see as many black nodes as we do when we go down from the root
to the same *a* to *e* in the second tree.

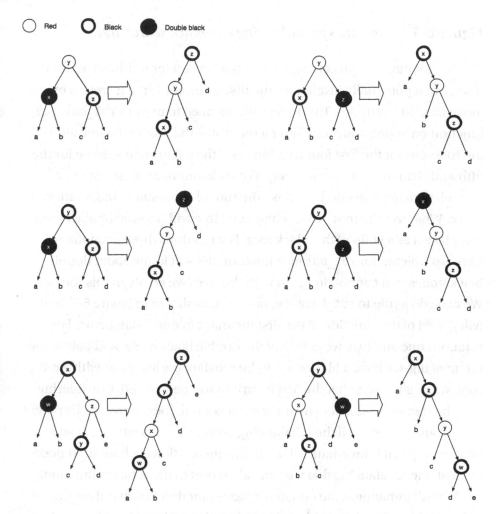

Figure 6-4. *Rotating sub-trees in red-black search trees*

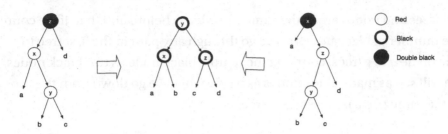

Figure 6-5. *Extra rules for balancing red-black search trees*

The rotations all preserve the second invariant for red-black search trees, but it potentially invalidates the first invariant when it creates or move around red nodes. To correct this, we need to use our `rbt_balance` function on appropriate trees after a rotation—the root of the tree after the rotations for the first four rotations and the left or right subtree for the fifth and sixth rotation, respectively. For rotations five and six, at most one rebalancing is needed, because the root of the result of the rotation is black. Whatever the root of the rebalanced tree will be—red or black—we can place it as a child of this black root. For rotations three and four, we have a problem: our `rbt_balance` function doesn't know about double black nodes so it cannot do much with the trees we create in this rotation. We can add a rule to `rbt_balance`, however, as shown in Figure 6-5, that will get rid of the violation of the first invariant in one rebalancing. For rotations one and two, we get rid of the double black node, and before the rebalancing, we have a black root. If the rebalancing leaves us with a red root, we might worry that the first invariant will be violated. Considering that the tree we have *before* the rotation has a red node, however, its parent cannot also be red, so if the rebalancing gives us a red root, we are still conforming to the invariant. All in all, this means that each rotation needs at most one rebalancing that we can call as part of the rotation function.

We will implement the updated `remove` function first and then the `rbt_rotate` and updated `rbt_balance` functions. The `remove` function looks like this:

```r
remove.red_black_tree <- function(x, elm, ...) {
  new_tree <- rbt_remove(x, elm)
  new_tree$colour <- BLACK
  new_tree
}
```

with the real work being done by the helper function rbt_remove:

```r
rbt_remove <- function(tree, elm) {
  if (is_empty(tree)) { # we didn't find the value...
    return(tree)
  }

  if (tree$value == elm) { # found the value to delete
    a <- tree$left
    b <- tree$right
    if (is_empty(a) && is_empty(b)) { # leaf
      if (tree$colour == BLACK)
        return(red_black_tree_node(DOUBLE_BLACK, NA, NULL,
        NULL))
      else
        return(red_black_tree_node(BLACK, NA, NULL, NULL))

    } else if (is_empty(a) || is_empty(b)) { # one empty child
      non_empty <- if (is_empty(a)) b else a
      non_empty$colour <- BLACK
      return(non_empty)

    } else { # inner node
      s <- st_leftmost(tree$right)
      return(rbt_rotate(tree$colour, s, a, rbt_remove(b, s)))
    }
  }
}
```

```
# we need to search further down to remove the element
if (elm < tree$value)
  rbt_rotate(tree$colour, tree$value,
             rbt_remove(tree$left, elm), tree$right)
else # (elm > tree$value)
  rbt_rotate(tree$colour, tree$value,
             tree$left, rbt_remove(tree$right, elm))
}
```

Compared to the unbalanced st_remove, the difference is the re-coloring we do when we find a case with one or two empty children and that we wrap recursive rbt_remove calls in a rbt_rotate call.

The real work is being done by the rbt_rotate function that implements the transitions from Figure 6-4. This is done with a series of pattern matchings and transformations:

```
rbt_rotate <- function(colour, value, left, right) {
  # first case
  if (pattern_match(a_x_b = left, c = right$left,
  d = right$right,
                    y = value, z = right$value,
                    a_x_b$colour == DOUBLE_BLACK,
                    colour == RED,
                    right$colour == BLACK)) {

    a_x_b$colour <- BLACK
    rbt_balance(BLACK, z,
                red_black_tree_node(RED, y, a_x_b, c),
                d)

  } else if (pattern_match(a = left$left, b = left$right,
                           c_z_d = right,
                           y = value, x = left$value,
                           left$colour == BLACK,
```

```
                                colour == RED,
                                c_z_d$colour == DOUBLE_BLACK)) {

  c_z_d$colour <- BLACK
  rbt_balance(BLACK, x,
                a,
                red_black_tree_node(RED, y, b, c_z_d))

# second case
} else if (pattern_match(a_x_b = left, c = right$left,
d = right$right,
                          y = value, z = right$value,
                          colour == BLACK,
                          a_x_b$colour == DOUBLE_BLACK,
                          right$colour == BLACK)) {

  a_x_b$colour <- BLACK
  new_left <- red_black_tree_node(RED, y, a_x_b, c)
  rbt_balance(DOUBLE_BLACK, z, new_left, d)

} else if (pattern_match(a = left$left, b = left$right,
                          y = value, c_z_d = right,
                          left$colour == BLACK,
                          colour == BLACK,
                          c_z_d$colour == DOUBLE_BLACK)) {

  c_z_d$colour <- BLACK
  new_right <- red_black_tree_node(RED, y, b, c_z_d)
  rbt_balance(DOUBLE_BLACK, x, a, new_right)

# third case
} else if (pattern_match(a_w_b = left,
                          c = right$left$left,
                          d = right$left$right,
                          e = right$right,
```

```
                                x = value, z = right$value,
                                y = right$left$value,
                                a_w_b$colour == DOUBLE_BLACK,
                                colour == BLACK,
                                right$colour == RED)) {

  a_w_b$colour <- BLACK
  new_left_left <- red_black_tree_node(RED, x, a_w_b, c)
  new_left <- rbt_balance(BLACK, y, new_left_left, d)
  red_black_tree_node(BLACK, z, new_left, e)

} else if (pattern_match(a = left$left,
                         b = left$right$left,
                         c = left$right$right,
                         d_w_e = right,
                         z = left$right$value,
                         x = left$value,
                         y = value,
                         left$colour == RED,
                         colour == BLACK,
                         d_w_e$colour == DOUBLE_BLACK)) {

  d_w_e$colour <- BLACK
  new_right_right <- red_black_tree_node(RED, y, c, d_w_e)
  new_right <- rbt_balance(BLACK, z, b, new_right_right)
  red_black_tree_node(BLACK, x, a, new_right)

} else {
  red_black_tree_node(colour, value, left, right)
}
}
```

Notice the use of rbt_balance to construct some of the trees. This will fix any violations of the first invariant and is called on the trees where this invariant might be invalidated. The updated version of rbt_balance, which includes the rules from Figure 6-5, looks like this:

```
rbt_balance <- function(colour, value, left, right) {
  if (pattern_match(a = left$left,
                    b = left$right$left,
                    c = left$right$right,
                    d = right,
                    x = left$value,
                    y = left$right$value,
                    z = value,
                    colour == BLACK,
                    left$colour == RED,
                    left$right$colour == RED)

    || pattern_match(a = left$left$left,
                     b = left$left$right,
                     c = left$right,
                     d = right,
                     x = left$left$value,
                     y = left$value,
                     z = value,
                     colour == BLACK,
                     left$colour == RED,
                     left$left$colour == RED)

    || pattern_match(a = left,
                     b = right$left,
                     c = right$right$left,
                     d = right$right$right,
                     x = value,
                     y = right$value,
                     z = right$right$value,
```

```
                      colour == BLACK,
                      right$colour == RED,
                      right$right$colour == RED)
    || pattern_match(a = left,
                     b = right$left$left,
                     c = right$left$right,
                     d = right$right,
                     x = value,
                     y = right$left$value,
                     z = right$value,
                     colour == BLACK,
                     right$colour == RED,
                     right$left$colour == RED)
) {

  left <- red_black_tree_node(BLACK, x, a, b)
  right <- red_black_tree_node(BLACK, z, c, d)
  red_black_tree_node(colour = RED, value = y, left, right)

} else if (pattern_match(a = left$left,
                        b = left$right$left,
                        c = left$right$right,
                        d = right,
                        z = value,
                        x = left$value,
                        y = left$right$value,
                        colour == DOUBLE_BLACK,
                        left$colour == RED,
                        left$right$colour == RED)

          || pattern_match(a = left,
                          b = right$left$left,
```

```
                          c = right$left$right,
                          d = right$right,
                          x = value,
                          z = right$value,
                          y = right$left$value,
                          colour == DOUBLE_BLACK,
                          right$colour == RED,
                          right$left$colour == RED)) {

    left <- red_black_tree_node(BLACK, x, a, b)
    right <- red_black_tree_node(BLACK, z, c, d)
    red_black_tree_node(BLACK, y, left, right)

  } else {
    red_black_tree_node(colour, value, left, right)
  }
}
```

To see how the balancing affects the search trees, we can calculate the minimal and maximal depth for red-black trees and compare with unbalanced trees. The following function recursively computes the minimal and maximal depth of nodes in a tree:

```
min_max_depth <- function(tree) {
  if (is_empty(tree))
    return(c(0, 0))
  left_min_max <- min_max_depth(tree$left)
  right_min_max <- min_max_depth(tree$right)
  min_depth <- min(left_min_max[1], right_min_max[1]) + 1
  max_depth <- max(left_min_max[2], right_min_max[2]) + 1
  c(min = min_depth, max = max_depth)
}
```

We can use it in an experimental setup where we can process the numbers from 1 to *n*—we will either just leave them ordered or give them a random permutation—and then build a tree using the two different search tree structures for a number of *n*s and collect them in a data frame:

```
library(tibble)
measure_depth <- function(process, empty_tree, ns) {
  measure_depth <- function(n) {
    tree <- empty_tree
    for (x in process(1:n))
      tree <- insert(tree, x)
    c(n = n, min_max_depth(tree))
  }
  measure_depth <- Vectorize(measure_depth)
  measurements <- measure_depth(ns)
  as_tibble(t(measurements))
}
```

For the ordered list, the experiment is deterministic, but for the random permutation we will run the experiment ten times to see the variation:

```
ns <- seq(1, 50, by = 5)
ns_rep <- rep(ns, times = 10)
depth <- rbind(
  cbind(data_structure = "Unbalanced",
        data = "Ordered",
        measure_depth(identity, empty_search_tree(), ns)),
  cbind(data_structure = "Red-black",
        data = "Ordered",
        measure_depth(identity, empty_red_black_tree(), ns)),
  cbind(data_structure = "Unbalanced",
        data = "Random",
```

```
      measure_depth(sample, empty_search_tree(), ns_rep)),
  cbind(data_structure = "Red-black",
        data = "Random",
        measure_depth(sample, empty_red_black_tree(), ns_rep))
)
```

We can then plot the result:

```
library(ggplot2)
ggplot(depth, aes(x = n, colour = data_structure)) +
  geom_jitter(aes(y = min)) +
  geom_smooth(aes(y = min), method = "loess",
                span = 2, se = FALSE) +
  geom_jitter(aes(y = max)) +
  geom_smooth(aes(y = max), method = "loess",
                span = 2, se = FALSE) +
  facet_grid(. ~ data) +
  scale_colour_grey("Data structure", end = 0.5) +
  xlab(quote(n)) + ylab("Tree depth") +
  theme_minimal() +
  theme(axis.text.x = element_text(angle = 90, hjust = 1))
```

The plot is shown in Figure 6-6. When we insert the elements in increasing order, the unbalanced tree will be maximally unbalanced. The maximal depth is one or two elements from n, whereas the minimal depth is one or two. For the red-black tree, the distance from the minimal to the maximal depth is guaranteed to be no more than a factor of two. If the elements are inserted in random order, the "unbalanced" tree is actually balanced—not quite as balanced as the red-black tree, but the difference between the maximal and minimal depth does not grow linearly.

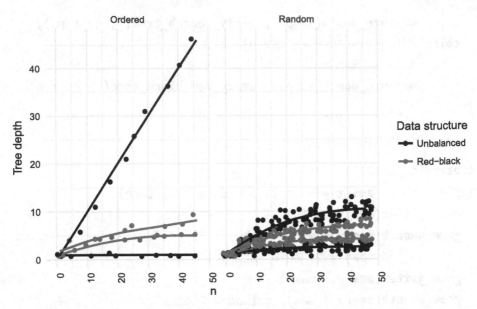

Figure 6-6. *Search tree depth for balanced (red-black) and unbalanced trees*

We can then explore how this affects the time it takes to construct a search tree. To run the time experiments for tree construction, we can use these functions:

```
setup_for_construction <- function(n) n
evaluate_construction <- function(empty, process)
function(n, x) {
  tree <- empty
  elements <- process(1:n)
  for (elm in elements) {
    tree <- insert(tree, elm)
  }
  tree
}
```

In these experiments, we want to vary both the data and the data structure, so we modify the performance measurement functions slightly to include data information:

```
get_performance_n <- function(
  algo
  , data
  , n
  , setup
  , evaluate
  , times
  , ...) {

  config <- setup(n)
  benchmarks <- microbenchmark(evaluate(n, config),
  times = times)
  tibble(algo = algo, data = data,
       n = n, time = benchmarks$time / 1e9)
}

get_performance <- function(
  algo
  , data
  , ns
  , setup
  , evaluate
  , times = 10
  , ...) {
  f <- function(n)
    get_performance_n(algo, data, n,
                      setup, evaluate,
                      times = times, ...)
```

```
  results <- Map(f, ns)
  do.call('rbind', results)
}
```

For the unbalanced tree, the insertion functions will run out of stack space for larger trees, so we split the experiments into small size and medium size trees and only construct unbalanced trees for ordered input for the small sizes:

```
ns <- seq(100, 450, by = 50)
performance_small <- rbind(
  get_performance("Unbalanced", "Increasing order", ns,
                  setup_for_construction,
                  evaluate_construction(empty_search_tree(),
                  identity)),
  get_performance("Unbalanced", "Random order", ns,
                  setup_for_construction,
                  evaluate_construction(empty_search_tree(),
                  sample)),
  get_performance("Red-black", "Increasing order", ns,
                  setup_for_construction,
                  evaluate_construction(empty_red_black_tree(),
                  identity)),
  get_performance("Red-black", "Random order", ns,
                  setup_for_construction,
                  evaluate_construction(empty_red_black_tree(),
                  sample))
)

ns <- seq(500, 1000, by = 50)
performance_medium <- rbind(
  get_performance("Unbalanced", "Random order", ns,
                  setup_for_construction,
```

```
            evaluate_construction(empty_search_tree(),
            sample)),
  get_performance("Red-black", "Increasing order", ns,
            setup_for_construction,
            evaluate_construction(empty_red_black_tree(),
            identity)),
  get_performance("Red-black", "Random order", ns,
            setup_for_construction,
            evaluate_construction(empty_red_black_tree(),
            sample))
)
```

To construct a tree of size n, we would expect the time to be $O(n^2)$ if the tree is unbalanced and $O(n\log n)$ if it is. We will plot the time divided by $O(n\log n)$; if the trees are balanced, the plots should flatten out as a line:

```
performance <- rbind(performance_small, performance_medium)
ggplot(performance,
       aes(x = n, y = time / (n*log(n)), colour = algo)) +
  geom_jitter() +
  geom_smooth(method = "loess",
                span = 2, se = FALSE) +
  facet_grid(. ~ data) +
  scale_colour_grey("Data structure", end = 0.5) +
  xlab(quote(n)) + ylab(expression(Time / n*log(n))) +
  theme_minimal() +
  theme(axis.text.x = element_text(angle = 90, hjust = 1))
```

The result is shown in Figure 6-7. As we would expect, the unbalanced tree quickly gets slow when the input is given in increasing order, whereas the red-black tree has roughly the same running time for ordered and random input. This, however, comes at a cost: as we can clearly see, for the randomly ordered input, the unbalanced tree, which doesn't have any transformation rules, is *much* faster if the input just happens to make it balanced. If possible, randomizing the input seems to be a better strategy than balancing the tree during construction. This, incidentally, is not that uncommon when it comes to algorithms. One of the fastest sorting algorithms around, quick-sort, has worst-case $O(n^2)$ but expected case $O(n\log n)$ running time, and outperforms algorithms with $O(n\log n)$ worst-case performance by just being simpler.

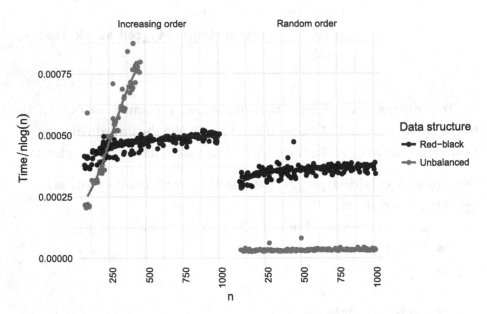

Figure 6-7. *Construction time for balanced and unbalanced trees*

We can also validate the membership test running time. The two implementations use exactly the same membership test function, but the running time will depend on the depth of the trees. We can set up an experiment where we build trees of different sizes and in all experiments measure how long it takes to search for a constant number of elements— we will use 100 elements:

```
setup_for_member <- function(empty, process) function(n) {
  tree <- empty
  elements <- process(1:n)
  for (elm in elements) {
    tree <- insert(tree, elm)
  }
  tree
}
evaluate_member <- function(n, tree) {
  elements <- sample(1:n, size = 100, replace = TRUE)
  for (elm in elements) {
    member(tree, elm)
  }
}
```

Again, I have split the experiments in two to avoid too large unbalanced trees (the evaluation code is similar to the construction case and is not shown). The result is shown in Figure 6-8, where the full experiment is shown on the right while I have zoomed in on the fast operations on the left. We clearly see that the unbalanced tree, in its worst-case scenario, where it is built on elements in increasing order, is dramatically slower than the balanced trees, but we also see that the slightly better balance of red-black search trees means that these outperform the simpler tree for membership queries. The extra complexity in red-black trees is irrelevant when it comes to member queries—the two trees use exactly the same function—it only affects insertion and deletion.

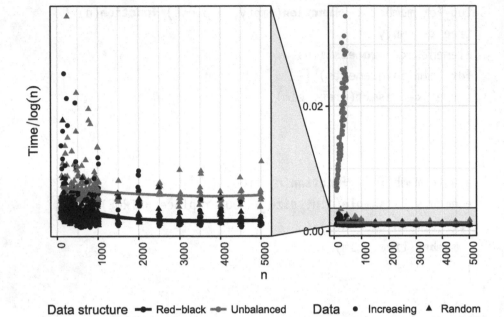

Data structure ● Red–black ● Unbalanced   Data ● Increasing ▲ Random

Figure 6-8. *Membership running time for unbalanced and black-red search trees*

Speaking of deletion, we can also experiment with that operation. We set it up similar to the member performance experiments. We build a tree of size *n* and then delete 100 elements from it—which, of course, will leave an empty tree when *n* is less than 100. But that is okay—we are mostly interested in the larger trees anyway, where the balancedness is more important:

```
setup_for_remove <- function(empty, process) function(n) {
  tree <- empty
  elements <- process(1:n)
  for (elm in elements) {
    tree <- insert(tree, elm)
  }
  tree
}
```

224

```
evaluate_remove <- function(n, tree) {
  elements <- sample(1:n, size = 100, replace = TRUE)
  for (elm in elements) {
    tree <- remove(tree, elm)
  }
}
```

Results of the experiment are shown in Figure 6-9. The results indicate, similar to the construction results, that the worst-case performance for the unbalanced tree is abysmal. We cannot even work with trees larger than about 500 elements without running into recursion depth problems. However, the simpler tree achieves a much better performance on average for trees built on data in random order.

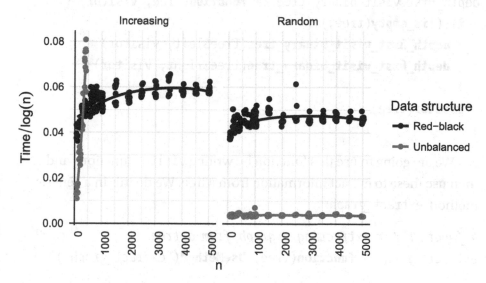

Figure 6-9. *Removal of elements in search trees*

Visualizing Red-Black Trees

For visualizing red-black trees, we could write a plot.red_black_tree
function similar to the plot.unbalanced_search_tree function we wrote
in Chapter 3, but this would duplicate a lot of code. Instead, we can exploit
that the two search trees now share an abstract superclass, search_tree,
and we can move some of the functionality there. We can reuse the
node_number_annotate_tree function from Chapter 3 as it is, but we will
refactor the tree traversal into a function responsible for recursing over the
tree and another for extracting whatever information we need from each
node. The traversal function looks like this:

```
# helper function for traversing trees
depth_first_visit_binary_tree <- function(tree, visitor) {
  if (!is_empty(tree)) {
    depth_first_visit_binary_tree(tree$left, visitor)
    depth_first_visit_binary_tree(tree$right, visitor)
  }
  visitor(tree)
}
```

We are going to use this function by writing visitor functions and
then use these to extract information from a tree. We do this in a generic
method, extract_graph:

```
# function for extracting a graph from a tree
extract_graph <- function(tree) UseMethod("extract_graph")
```

The extract_graph for the search_tree class extracts the basic tree
topology and the values in the non-empty nodes:

```
extract_graph.search_tree <- function(tree) {
  n <- tree$dfn
  values <- vector("numeric", length = n)
  from <- vector("integer", length = n - 1)
```

```
  to <- vector("integer", length = n - 1)
  edge_idx <- 1

  extract <- function(tree) {
    # we change the index so the root is number 1
    i <- n - tree$dfn + 1
    values[i] <<- ifelse(is.na(tree$value), "", tree$value)

    if (!is_empty(tree)) {
      j <- n - tree$left$dfn + 1
      from[edge_idx] <<- i
      to[edge_idx] <<- j
      edge_idx <<- edge_idx + 1

      k <- n - tree$right$dfn + 1
      from[edge_idx] <<- i
      to[edge_idx] <<- k
      edge_idx <<- edge_idx + 1
    }
  }

  depth_first_visit_binary_tree(tree, extract)
  nodes <- tibble(value = values)
  edges <- tibble(from = from, to = to)
  list(nodes = nodes, edges = edges)
}
```

This is enough to plot basic trees:

```
plot.search_tree <- function(x, ...) {
  x %>% node_number_annotate_tree %>%
    extract_graph %$% tbl_graph(nodes, edges) %>%
    mutate(leaf = node_is_leaf()) %>%
    ggraph(layout = "tree") +
```

```
    scale_x_reverse() +
    geom_edge_link() +
    geom_node_point(aes(filter = leaf),
                    size = 2, shape = 21,
                    fill = "black") +
    geom_node_point(aes(filter = !leaf),
                    size = 10, shape = 21,
                    fill = "white") +
    geom_node_text(aes(label = value),
                    vjust = 0.4) +
    theme_graph()
}
```

There is no additional meta-information in an unbalanced tree, so we can just let the unbalanced_search_tree inherit these methods.

For the red-black trees, though, we also want to show the colors in each node. So we need a specialization of extract_graph for extracting these and a specialization of plot for displaying them.

For extract_graph, we will call the superclass method to get the topology information and then extend the nodes tibble with color information:

```
extract_graph.red_black_tree <- function(tree) {
  n <- tree$dfn
  colours <- vector("numeric", length = n)
  extract <- function(tree) {
    # we change the index so the root is number 1
    i <- n - tree$dfn + 1
    colours[i] <<- tree$colour
  }
  depth_first_visit_binary_tree(tree, extract)

  graph <- NextMethod()
  RB <- c("Red", "Black", "Double black")
```

```
  nodes <- graph$nodes %>% add_column(colour = RB[colours])
  edges <- graph$edges
  list(nodes = nodes, edges = edges)
}
```

For plotting, we call the superclass method and then modify the plot before we return it:

```
plot.red_black_tree <- function(x, ...) {
  NextMethod() +
    scale_fill_manual("Colour",
                       values = c("Red" = "white",
                                  "Black" = "black",
                                  "Double black" = "lightgray")) +
    geom_node_point(aes(filter = leaf, fill = colour),
                    size = 2, shape = 21) +
    geom_node_point(aes(filter = !leaf, fill = colour),
                    size = 10, shape = 21) +
    geom_node_text(aes(filter = colour == "Black",
                       label = value),
                   colour = 'white', vjust = 0.4) +
    geom_node_text(aes(filter = colour == "Double black",
                       label = value),
                   colour = 'black', vjust = 0.4) +
    geom_node_text(aes(filter = colour == "Red",
                       label = value),
                   colour = 'black', vjust = 0.4)
}
```

The red-black tree plotting includes the possibility of double black nodes. We would never expect to *see* a double black node in a consistent tree—these should only exist until we have bubbled them away in deletion—but for debugging purposes, it is helpful to include it. I've chosen to plot the black nodes as black—in the transformation figures I have used a bolder stroke instead to reserve black for double black—and of course we could plot the red nodes as red instead of white, but since this book is printed in black and white, I have stuck with white.

We can plot a tree like this:

```
rb_tree <- empty_red_black_tree()
for (i in 1:10)
  rb_tree <- insert(rb_tree, i)
plot(rb_tree)
```

The result is shown in Figure 6-10.

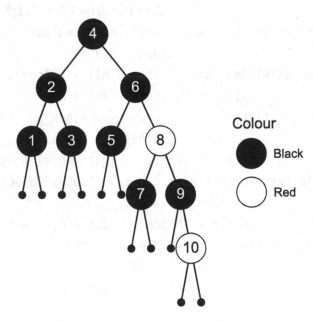

Figure 6-10. *Red-black search tree plotted in R*

Splay Trees

The splay tree structure we used for implementing a heap in the previous chapter is really a search tree, and if we implement the search tree operations, we can use it as such. We need a few modifications to get it to work as a search tree, though.

With splay trees, the underlying idea is that we modify the tree whenever we search for a member. We only did this when deleting the minimal element in the heap version, but when using a splay tree as a search tree, we also want to do this whenever we call member. This is problematic because we want member to return a boolean value and not a modified tree, so we cannot make a purely functional version of splay trees and at the same time have the same interface to the tree via the member function. We *could* make a member function that returns a modified tree, but instead we are going to break the purely functional approach and modify the tree when we access it. This makes the data structure ephemeral, but if you need it to be persistent, you can change the member function to return a modified tree instead and work from there. I trust that you will be able to do this after we have seen how to make an ephemeral version of the data structure.

The way that splay trees work as a search tree is that whenever we call member on a value, we will move that value up to the root of the tree. We do this with pattern matching, where we recognise certain tree structures and rearrange the tree for these. These rearrangements are called the splay operation, and at the end of splaying the element we search for in member will be the root, if it is in the tree. After that, we can check membership by looking at the value in the root of the tree. To be able to modify the tree as a side effect of membership queries, we store the tree in an environment—this is the only way we can have side effects in a data structure, after all. We implement the data structure like this:

```
empty_splay_tree <- function() {
  ref <- new.env(parent = emptyenv())
  ref$tree <- empty_splay_node()
  structure(ref, class = c("splay_tree", "search_tree"))
}

is_empty.splay_tree <- function(x)
  is_empty(x$tree)
```

This reuses the structure and emptiness test we implemented for the splay heap.

Inserting elements into a splay tree is the same operation as for the splay heap. We partition the existing elements into those that are smaller than the new elements and those that are larger, and then we insert the new element as the root of a tree that has the smaller elements as its left subtree and the larger elements as the right subtree. For this, we reuse the partition function from the splay heap. We have to modify the insert function in one way, though. For heaps, we allow the tree to contain more than one copy of the same value—this is necessary if we want to use a heap for sorting, after all. For search trees, we want to treat the structure as a set, so we do not want to insert an element if it is already in the tree. We can handle this by checking for membership before we insert an element. This has the added benefit that it restructures the tree to have the element, or an element close to it, at the root, which speeds up the partitioning, but that is a minor detail that doesn't change the functionality. The insert function then looks like this:

```
insert.splay_tree <- function(x, elm) {
  if (member(x, elm))
    return(x) # don't insert if we already have the element
```

```
  part <- partition(elm, x$tree)
  x$tree <- splay_tree_node(
    value = elm,
    left = part$smaller,
    right = part$larger
  )
  x
}
```

When deleting elements, we can treat a splay tree as we did the unbalanced search tree. We *could* try to balance the tree as part of deletion, but there is no guarantee that this will improve performance in subsequent operations, so we won't bother. We will just use the same deletion logic as we did for the unbalanced tree. We need a new function for it, though, because we need to construct a splay tree, with the objects of the right class, in the recursions, but the logic is exactly the same as what we have already seen:

```
splay_remove <- function(tree, elm) {
  # if we reach an empty tree, there is nothing to do
  if (is_empty(tree)) return(tree)

  if (tree$value == elm) {
    a <- tree$left
    b <- tree$right
    if (is_empty(a)) return(b)
    if (is_empty(b)) return(a)

    s <- st_leftmost(tree$right)
    return(splay_tree_node(s, a, splay_remove(b, s)))
  }

  # we need to search further down to remove the element
  if (elm < tree$value)
    splay_tree_node(tree$value, splay_remove(tree$left, elm),
    tree$right)
```

```
  else # (elm > tree$value)
    splay_tree_node(tree$value, tree$left, splay_remove
    (tree$right, elm))
}

remove.splay_tree <- function(x, elm) {
  x$tree <- splay_remove(x$tree, elm)
  x
}
```

It is in the member function that the splay magic happens that balances the tree. Whenever we test for membership, we first splay the tree, which will move the element we are looking for to the root if it is in the tree; after that, testing membership is a simpler matter of looking at the value in the root:

```
member.splay_tree <- function(x, elm) {
  x$tree <- splay(x$tree, elm)
  # if elm is in the tree it is now at the root
  !is.na(x$tree$value) && x$tree$value == elm
}
```

This leaves the splay function. This function should move the element we are looking for to the root and balance the tree at the same time. The operations involved in splaying are shown in Figure 6-11, and the full implementation of the function is listed next. As with so many other tree operations, it is mainly a case of pattern matching the tree structure, and for that, we reuse the pattern_match function from earlier. The different operations look at different depths into the tree. If we were just aiming to rotate the element we splay up to the root of the tree, we could do that with just the "Zig" and "Zag" operations and only look at one level of the tree. But that would not balance the tree to the same degree as when the other operations are also used, and would thus not achieve an expected amortized logarithmic depth of the tree.

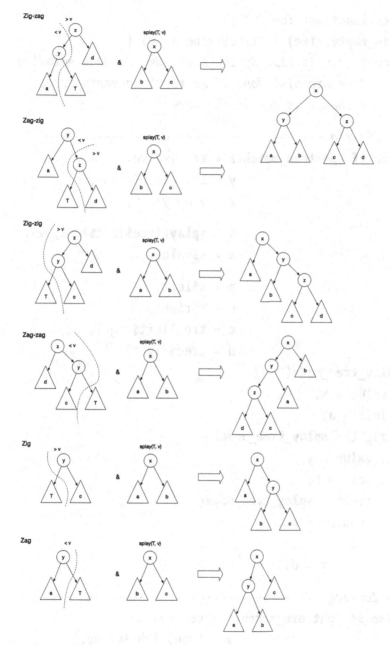

Figure 6-11. *Splay operations*

```
splay <- function(tree, v) {
  if (is_empty(tree) || tree$value == v) {
    tree # if v is already the root, we are done splaying
         # we are also done if we reach an empty tree;
         # then v is not in the tree

  # -- Zig-zig ------------------------------------
  } else if (pattern_match(z = tree$value,
                           y = tree$left$value,
                           z > v && y > v,

                           s = splay(tree$left$left, v),
                           x = s$value,

                           a = s$left,
                           b = s$right,
                           c = tree$left$right,
                           d = tree$right)) {
    splay_tree_node(
      value = x,
      left = a,
      right = splay_tree_node(
        value = y,
        left = b,
        right = splay_tree_node(
          value = z,
          left = c,
          right = d)))

  # -- Zag-zag ------------------------------------
  } else if (pattern_match(z = tree$value,
                           y = tree$right$value,
                           z < v && y < v,
```

```
                               s = splay(tree$right$right, v),
                               x = s$value,

                               a = s$left,
                               b = s$right,
                               c = tree$right$left,
                               d = tree$left)) {
    splay_tree_node(
      value = x,
      left = splay_tree_node(
        value = y,
        left = splay_tree_node(
          value = z,
          left = d,
          right = c),
        right = a),
      right = b)

# -- Zig-zag & zag-zig ------------------------
} else if (pattern_match(z = tree$value,
                         y = tree$left$value,
                         v < z && v > y,

                         s = splay(tree$left$right, v),
                         x = s$value,

                         a = tree$left$left,
                         b = s$left,
                         c = s$right,
                         d = tree$right)
           ||
           pattern_match(y = tree$value,
                         z = tree$right$value,
```

```r
                                  y < v && z > v,

                                  s = splay(tree$right$left, v),
                                  x = s$value,

                                  a = tree$left,
                                  b = s$left,
                                  c = s$right,
                                  d = tree$right$right)) {
  splay_tree_node(
      value = x,
      left = splay_tree_node(
        value = y,
        left = a,
        right = b),
      right = splay_tree_node(
        value = z,
        left = c,
        right = d))

  # -- Zig -------------------------------------
  } else if (pattern_match(y = tree$value,
                                  y > v,

                                  s = splay(tree$left, v),
                                  x = s$value,

                                  a = s$left,
                                  b = s$right,
                                  c = tree$right)) {
    splay_tree_node(
      value = x,
      left = a,
      right = splay_tree_node(
```

```
        value = y,
        left = b,
        right = c))
# -- Zag -------------------------------------
} else if (pattern_match(y = tree$value,

                         y < v,

                         s = splay(tree$right, v),
                         x = s$value,

                         a = tree$left,
                         b = s$left,
                         c = s$right)) {
  splay_tree_node(
    value = x,
    left = splay_tree_node(
      value = y,
      left = a,
      right = b),
    right = c)

} else {
  # if the recursive splay operation returns an empty tree,
  # which can happen if v is not in the tree, we reach this point
  # and here we just give up and return the tree.
  tree

}
}
```

For completeness, we can implement this plot function. This is very simple because we have already implemented plotting for splay trees for the heap (although I have not listed the implementation in this book, you can find it at http://github.com/mailund/ralgo), so we can reuse that implementation:

```
plot.splay_tree <- function(x) {
  plot(x$tree)
}
```

We can use this plotting function to explore how the balancing behaves when we call the member function on splay trees. If we run the following code, we will get the trees shown in Figure 6-12, with A showing the tree after we have built it from numbers 1 to 10 in order, B the result of querying for 4, C for 8, and D for 6. Notice how the element we query for is moved to the root of the tree, but also that elements we have recently queried for remain close to the root. It is not uncommon with search patterns where a small subset of elements are queried for repeatedly for a while before moving to another subset that is queried for. In such cases, the splay tree's balancing moves the subset towards the root of the tree and makes the processing faster than it would be if we didn't adapt the tree to the actual queries.

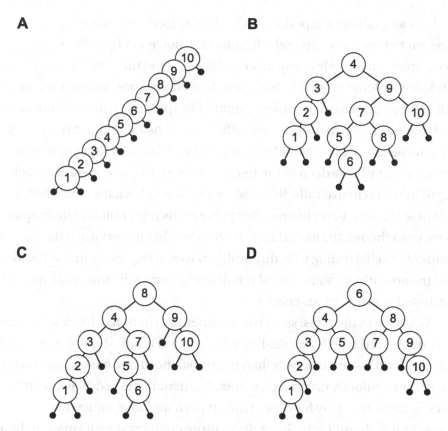

A   **B**   **C**   **D**

Figure 6-12. *Splay trees after balancing*

```
tree <- empty_splay_tree()
for (x in 1:10)
  tree <- insert(tree, x)
plot(tree)

member(tree, 4)
plot(tree)

member(tree, 8)
plot(tree)

member(tree, 6)
plot(tree)
```

We can evaluate the performance of splay trees by reusing the construction experiments code. Results are shown in Figure 6-13. For larger trees, I have only constructed splay trees and unbalanced trees for randomly ordered input. For both unbalanced trees and splay trees, we will get very deep trees for ordered input. For splay trees, this will not be a problem for performance—we can efficiently construct them even if they are unordered—but we would not be able to do membership queries for elements found very deep in the tree because we would exceed the stack depth in the recursive calls. Because we explicitly check for membership before we insert a new element, this prevents us from building deep splay trees, even though the partition function would not perform a deep recursion itself. Dealing with this problem would require us to use thunks and trampolines, or some sort of rebalancing during the insertion, as discussed in the previous chapter.

As we see in the plots, splay trees outperform both red-black trees and the unbalanced trees when dealing with ordered input—for the sizes that it can deal with it—but is slower than the unbalanced tree when constructing trees from randomly ordered elements. It outcompetes red-black search trees because the splay function, while it perhaps looks somewhat complicated, doesn't actually perform more complex transformation than the red-black search tree insertion function, and at the same time the tree nodes are simpler, and there are fewer values that need to be set. It is only slightly slower on randomly ordered data than it is for ordered data, which is because we first splay elements in the member call in both cases, but when the input data is randomized, the partition function needs to recurse deeper than when the data is input in order. The member and partition calls make insert on splay trees slightly more complex than the insert function on unbalanced trees, which is why it is slower there.

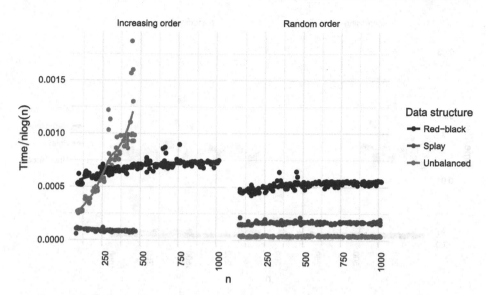

Figure 6-13. *Comparison of search tree construction*

For membership tests we can, as we did before, measure the time it takes to search for 100 random elements in trees of different size. Results are shown in Figure 6-14. Because the splay tree balances the tree in the member check when inserting elements, it performs better than the unbalanced tree on ordered input. But because the splay operation that is part of the membership test is much more complex than the simple recursive search performed by both unbalanced trees and red-black trees, membership checks are substantially slower when the other trees are balanced, as the red-black tree always is and the "unbalanced" tree is when the input is given to the construction method in random order.

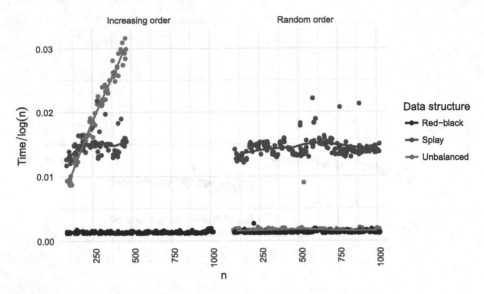

Figure 6-14. *Comparison of member tests in search trees*

If we perform repeated membership calls of the same few elements, using this function for the performance evaluation

```
evaluate_member <- function(n, tree) {
  elements <- rep(sample(1:n, size = 10, replace = TRUE), 100)
  for (elm in elements) {
    member(tree, elm)
  }
}
```

we do see an improvement in the performance of membership tests, as shown in Figure 6-15, but the overhead involved in the splay function still makes membership slower for splay trees than the two other trees. Which tree is best for any given application thus depends on both the order we can insert elements in (if we can randomize them, the "unbalanced" tree will be balanced, and it will outperform the other two) and how large the tree we build is versus how many membership tests we do—the splay tree is faster to build, but the red-black tree will be better for membership tests.

244

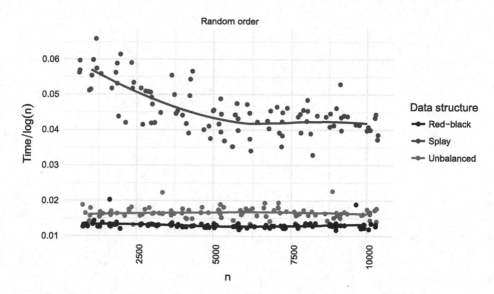

Figure 6-15. *Repeated membership tests for a small subset of elements*

Conclusions

R is not really a language for algorithmic programming. It is designed to be flexible and make statistical analysis and visualization easy, but not really to be fast. With modern compiler and runtime system technology, it could probably become much faster than it is today, but basic speed has not been a priority. *Programming* time—the time it takes you to implement an analysis—is prioritized higher than *running* time—the time it takes the computer to complete the analysis. This is the sensible choice for many applications. There is no point in spending an extra week programming to save a day of running time. Since programmers' time is more important than computers', it usually isn't worth it to program a week to save a *month* of running time.

How much extra time you should spend on making your code run faster is a trade-off against how long and how often you expect your code to run. If you are writing an analysis package you expect lots of people to use frequently, you should consider optimizing your code; if you are writing a one-off analysis pipeline, you probably shouldn't bother. Or, at least, you shouldn't spend too much time optimizing your code.

When it comes to efficient code, the greatest speed-ups are achieved by carefully selecting the algorithms and data structures you use. Changing a quadratic time algorithm into a linear time algorithm will significantly speed up your analyses for even moderately sized data sets; checking set membership in logarithmic time instead of linear time will do the same. If you find that your code is too slow for your purposes, it is the algorithms and data structures you should consider doing something about first. Micro-optimizing the actual code—painting it "go faster red" and putting speed stripes on it—might gain you a factor of two to ten in speed-up,

T. Mailund, *Functional Data Structures in R*, https://doi.org/10.1007/978-1-4842-3144-9

but changing an algorithm or data structure can gain you asymptotic improvements. When programming in R, micro-optimizing is rarely worth the effort, but switching from one data structure to another just might be. Especially if you interact with data structures through a polymorphic interface, changing one data structure for another requires trivial changes to the code but can give you great performance improvements.

R doesn't have an extensive library of data structures to choose from. Again, this is because performance has not been the driving force behind the development of the language. But we can build our own libraries, and we can make them reusable, and each time we improve a library, we make it easier for the next programmer to gain efficiency with minimal effort. I have done my little part by implementing the data structures described in this book available on GitHub at `https://github.com/mailund/ralgo`. Feel free to use my code in your analyses, and I would love to get pull requests with improvements.

This book does not provide an exhaustive list of functional data structures. There are many different variations of data structures, many of which can be translated into persistent/functional versions. I hope to see more of them in R packages in the future, and I hope this book has motivated you to try implementing some of them. If you do, I would love to hear about it.

Acknowledgements

I am grateful to Christian N. Storm Pedersen for fruitful discussions.

Bibliography

Brodal, G. S., and C. Okasaki. 1996. "Optimal Purely Functional Priority Queues." *Journal of Functional Programming* 6 (6): 839–57.

Crane, C. A. 1972. "Linear Lists and Priority Queues as Balanced Binary Trees." PhD thesis, Stanford, CA, USA: Stanford University.

Germane, K, and M Might. 2014. "Deletion: The curse of the red-black tree." *Journal of Functional Programming* 24 (04): 423–33.

Mailund, Thomas. 2017a. *Functional Programming in R: Advanced Statistical Programming for Data Science, Analysis and Finance*. Berkeley, CA: Apress.

Mailund, Thomas. 2017b. *Metaprogramming in R: Advanced Statistical Programming for Data Science, Analysis and Finance*. Berkeley, CA: Apress.

Mailund, Thomas. 2017c. *Advanced Object-Oriented Programming in R: Statistical Programming for Data Science, Analysis and Finance*. Berkeley, CA: Apress.

Okasaki, C. 1995a. "Purely Functional Random-Access Lists." In *In Functional Programming Languages and Computer Architecture*, 86–95. ACM Press.

Okasaki, C. 1995b. "Simple and Efficient Purely Functional Queues and Deques." *Journal of Functional Programming* 5 (4). Cambridge University Press: 583–92.

Okasaki, C. 1999a. *Purely Functional Data Structures*. Cambridge University Press.

T. Mailund, *Functional Data Structures in R*, https://doi.org/10.1007/978-1-4842-3144-9

Okasaki, C. 1999b. "Red-Black Trees in a Functional Setting." *Journal of Functional Programming* 9 (4). Cambridge University Press: 471–77.

Okasaki, C. 2005. "Alternatives to Two Classic Data Structures." *SIGCSE Bull.* 37 (1). New York, NY, USA: ACM: 162–65.

Wickham, H. 2014. *Advanced R*. Chapman & Hall/Crc the R Series. Taylor & Francis.

Index

© Thomas Mailund 2017
T. Mailund, *Functional Data Structures in R*, https://doi.org/10.1007/978-1-4842-3144-9

T, U, V, W, X, Y, Z

Get the eBook for only $5!

Why limit yourself?

With most of our titles available in both PDF and ePUB format, you can access your content wherever and however you wish—on your PC, phone, tablet, or reader.

Since you've purchased this print book, we are happy to offer you the eBook for just $5.

To learn more, go to http://www.apress.com/companion or contact support@apress.com.

Apress®

Printed in the United States
By Bookmasters